Sound, Space, and the City

THE CITY IN THE 21ST CENTURY

Eugenie L. Birch and Susan M. Wachter, Series Editors

A complete list of books in the series is
available from the publisher.

Sound, Space, and the City

Civic Performance
in Downtown Los Angeles

Marina Peterson

PENN

UNIVERSITY OF PENNSYLVANIA PRESS

PHILADELPHIA · OXFORD

Published by
University of Pennsylvania Press
Philadelphia, Pennsylvania 19104-4112

Printed in the United States of America on acid-free paper

10   9   8   7   6   5   4   3   2   1

Library of Congress Cataloging-in-Publication Data
ISBN 978-0-8122-4234-8

For my grandmothers,
Olive Eggan Redfield and Lois Elizabeth Morgan

Contents

Preface

I went to Los Angeles to see what happened in a city that did not take its centrality for granted. As my knowledge of Los Angeles at that time was acquired through the literature on the postmodern city and its features of sprawl and decentralization (Davis 1992; Dear 1986, 2000; Jencks 1993; Soja 1989, 1996), I had not expected to find a public concert presenter downtown. I was aware of Grand Performances and was in contact with its staff through the music programmer at the Chicago Cultural Center, where I had conducted research and performed (Peterson 2002). Inscribing a process of center-making, Chicago stands as a model for Angelenos invested in making downtown a center for the city in order to make L.A. a "real" city. With a park on the downtown lakefront and a radial organization of streets and neighborhoods, Chicago exemplifies "city" as defined by an urban form of centrality reified by the Chicago School of sociology (Park and Burgess 1925). Though naturalized as "urban," Chicago's centrality is continually made, from the initial implementation of the Burnham plan to the recent creation of Millennium Park.

Similarities between Grand Performances and the Chicago Cultural Center were apparent immediately. On the same world music circuit, they also share the primary intention of drawing audiences of diverse residents of their respective cities. Moreover, each of their downtown locations allows them to conceive of their own position as neutral. The differences between the two lie in their conceptions of the surrounding city, of the imagined "cityness" of their respective homes, and their institutional status as public concert presenters. In Chicago, supported by the city, free concerts are unquestioned by the presenters, described simply as "what we do." In Los Angeles, a nonprofit organization in a corporate plaza must consistently assert the significance of its work. Thus, though the existence of a downtown public concert presenter surprised me, its position in a neoliberal economy of privatization resulted in an institution that did not take itself for granted. Rather, the ways in which its centrality was continually being made were more apparent. Yet while this dynamic indicated the ways in which civic institutions and ideologies are produced, following the logic of the center, it occluded another center, that of the neoliberal state.

An anthropology of the center puts pressure on anthropological knowledge. Though the center is always constituted as such, it is defined by the fact that it masks its own making. As an anthropological subject, this poses problems for research and analysis. A center-periphery binary that has historically imbued anthropological inquiry tends to equate the center with emptiness and the periphery with sociality. While turning this structure inside out enables examination of certain social and political aspects of the center, it tends to disallow an assessment of the making of neutrality and a consideration of how that neutrality is part of the social meaning of a given case. Studying how the center is made as such requires paying attention to the construction and deployment of categories of centrality in practice, considering both the logics and ends of centrality as well as how those ends are recognized as having been achieved. (Though the contributions to this discussion are by now vast, some that have particularly informed my thinking are Appadurai 1996; Clifford and Marcus 1986; Fabian 1983; Gupta and Ferguson 1997; Marcus 2008; Rabinow 1995; Rabinow and Marcus 2008; and Trouillot 1991.)

In researching public concerts my informants were professionals, adept at framing their work through discourse that asserted its value. While our conceptual approaches and motivations may have differed, the material from which our conclusions were drawn was often the same. When ethnographic informants express what is also an analytic framework of anthropology, it reveals the situatedness of disciplinary knowledge and challenges the locations in which anthropological theory has been developed and applied. As such, it raises questions about what analytic frames are possible for these kinds of cases and, by extension, any cases. For instance, the articulation of a Durkheimian theory of ritual by public concert audience members, discussed in Chapter 5, suggests that ideas of social science seep into everyday life. Moreover, it indicates that those ideas come out of sociohistorical contexts, which, in this case, are also those of Los Angeles public concert audience members.

Conversely, insofar as multiculturalism is a project of Grand Performances, in my analysis I aim to refrain from inhabiting the logic of multiculturalism. Without taking a stand for or against multiculturalism, as much as possible I avoid using its categories of evaluation in describing or accounting for the organization, the performances, and the city, showing instead how the terms are used as part of a multicultural project. Hence I use information from census figures, Grand Performances' internal documents, and mailing list sign-ups in order to present demographics within their mode of construction. At the same time, labeling identity is sometimes unavoidable; my use of terms and capitalization reflects these shifts.

Finally, many of the photographs that might show a reader what the performance space looks like are those used by Grand Performances and its artists for their own publicity. I frame these as such, while acknowledging the slippage between our respective uses of the images. In the end, this slippage becomes a space of productive tension; the challenge is in taking it seriously while also locating an analytic position that is not the same as that of the subject of research. As these and other examples suggest, in the end there is no single outside framework or set of information. Instead, one moves within the warp and woof of a dynamic exchange between ethnographic material and theoretical framing.

With this project I aspire toward an anthropology *of* the city rather than *in* the city. Often necessarily multisited, urban anthropology relies on an experience of everyday life in the city, of being in the city at large while one's specific field site might be traveled to rather than resided in. Though an ethnographer's perspective is always partial, I am interested in how these downtown public performances are located in, connected to, and evoke the city of Los Angeles. In focusing on how meaning is made in and around public performances, I aim to engage with performance both in its own right and as integrated into the urban fabric. To this end I observed and participated in practices that constituted the imagining and making of a multicultural L.A., situating Grand Performances' work in contexts of historical and contemporary urban planning, artistic programming, and the city of Los Angeles as lived and imagined. Living downtown, I experienced urban dynamics of gentrification as the area was being transformed through loft development and an accompanying demographic shift. While events and openings brought a local artist community together, artists and homeless were at odds as new residents moved into rapidly developing loft spaces.

Ethnographic research was primarily conducted from 2001 to 2003; I have returned nearly every summer since and observe much of the same. Though the time period of research reflects a particular moment in Grand

Performances' history, it is one that articulates wider dynamics of cultural policy on municipal, state, and national levels, and of the politics of multiculturalism, international performance, and downtown development in Los Angeles that continue to shape these and other events and processes. Though the various subjects who help make public concerts possible might have had different conceptualizations of the processes by which the concert achieved certain effects, overall they expressed a shared belief in a multiculturalism that is part of wider social and political frameworks. Producers, performers, and audience each play their role in order for the concert to occur, and for it to be a site in which other meanings are generated. The concert helps produce those roles through the physical organization of space, with the audience organized as a collective in relation to but separated from the performer, and through the practices expected of all involved.

Public concerts are formal performances that require months of administrative and bureaucratic preparation. This work is also a means of performing the city, though the subjects mostly sit in an office, connecting to wider networks of programming, funding, and marketing by making phone calls and writing grants. Much of the ethnographic material drawn on herein is from work with Grand Performances as an institution and includes observation of practices, statements made informally or in interviews, and various ephemera and documentation. During performances I moved between roles, sometimes participating as an audience member, at other times as a researcher of audience members; often I worked alongside staff members and volunteers, and once I was onstage as a performer. These positions facilitated an understanding of the performances from different perspectives through an embodied subject position, each partial in itself yet contributing to the whole of what made the event.

Those times that I aimed to blend in as an audience member required that in order to capture the complexity and ephemerality of performance I supplemented field notes with video of the performers and audience; adding a step of representation and reification, using video as a tool for research also allowed for a finer grain of ethnographic writing and a richer analysis. In writing about learning to play the cello to sound like a drum machine, I relied on kinesthetic memory, reconstructing the experience later in my office to better describe it in words. Performing with the daKAH Hip Hop Orchestra allowed me to also sound the city musically; playing hip-hop parts on the cello, we were in the center of the event, with the fountain shooting behind us and an audience dancing amid the skyscrapers. Though the performers are the focus of the event, the audience—figured as representation and synecdoche of the city—is the locus for the recognition of diversity.

This project would not have been possible without the staff of Grand Performances. I am especially indebted to Michael Alexander, who not only allowed me to conduct research in and of Grand Performances but also included me in a wide range of events that, while salient for research, also provided me with an introduction to Los Angeles's cultural landscape. Leigh Ann Hahn, Dean Porter, Alice Platt, Karlee Decima, Fred Stites, Mark Baker, and Zindy Landeros went above and beyond in their generosity and openness and were always patient with my outsider status while including me in their activities. Board members also offered invaluable assistance. Of those who shared their time and memories, a few merit special thanks. Dan Rosenfeld allowed me to peruse his scrapbooks that he kept while working for the winning development firm for what would become California Plaza, and Peggy Adams loaned me documents pertaining to initial legal agreements for the Plaza. Two other board members gave permission to use their drawings in Chapter 5.

Artists made the performances happen, and I have drawn from their work with and without their knowledge. Lonnie Marshall, David Rojas, and El Vez generously allowed me to print their lyrics in this book. Geoff "Double G" Gallegos gave me the opportunity to perform at California Plaza and to have a great time doing so. Audience members were gracious in responding to questions at events and agreeing to extended phone interviews. Though all granted permission for their names to be used, I do so for the most part only in the case of public statements; instead I have largely chosen to identify people by titles or to keep them anonymous. I am grateful to those who let me use their words; I apologize for any slight I may have made and to anyone I have forgotten to thank here by name: this book is for everyone whose practices are inscribed in this ethnography.

The Department of Anthropology at the University of Chicago provided an intellectual space that allowed me to first conceive of this project. With intellectual generosity, Jean Comaroff, Arjun Appadurai, Elizabeth Povinelli, Saskia Sassen, and Martin Stokes supported a project that does not fit neatly within disciplinary boundaries. I have since had the benefit of input from colleagues near and far. Much of the material was presented at conferences, including those of the American Anthropological Association, the American Studies Association, and the International Association for the Study of Popular Music. At Ohio University, faculty and students in the School of Interdisciplinary Arts have contributed to the development of the project directly and indirectly. Members of the Urban Reading Group and a faculty writing group gave insightful advice for several chapters. Mitchell Goodman prepared the images and Margaret Pearce and the Ohio University Cartography Center made maps. Others who, as readers, interlocutors, sounding boards, and supportive friends have most notably helped shape the book are Vicki Brennan, Andrea Frohne, Gary McDonogh, Daniel Monterescu, Keith Murphy, Alessandra Raengo, Sarah Schrank, Jesse Shipley, and Hamza Walker. Peter Agree, at the University of Pennsylvania Press, has been a wonderful editor, ever available and alternately patient and pushing at just the right times. The spirited suggestions of two anonymous reviewers were key in crafting the manuscript.

My family is in this book in ways more and less apparent. My parents, Joanna Redfield Vaughn and John Peterson, put me on a path that could incorporate ideas, music, and social worlds. My brother Jesse is a wonderful friend and musical collaborator. At the last hour, on an afternoon in Missouri, my father and brothers, Jesse and Asa, came up with possible titles while my sisters, Alexandria and Anastasia, perused family photos inside my grandmother's house. My grandparents and great-uncle instilled in me the excitement of intellectual engagement and a curiosity about the world and gave me an early glimpse of what anthropology might be; I am honored to carry on that lineage.

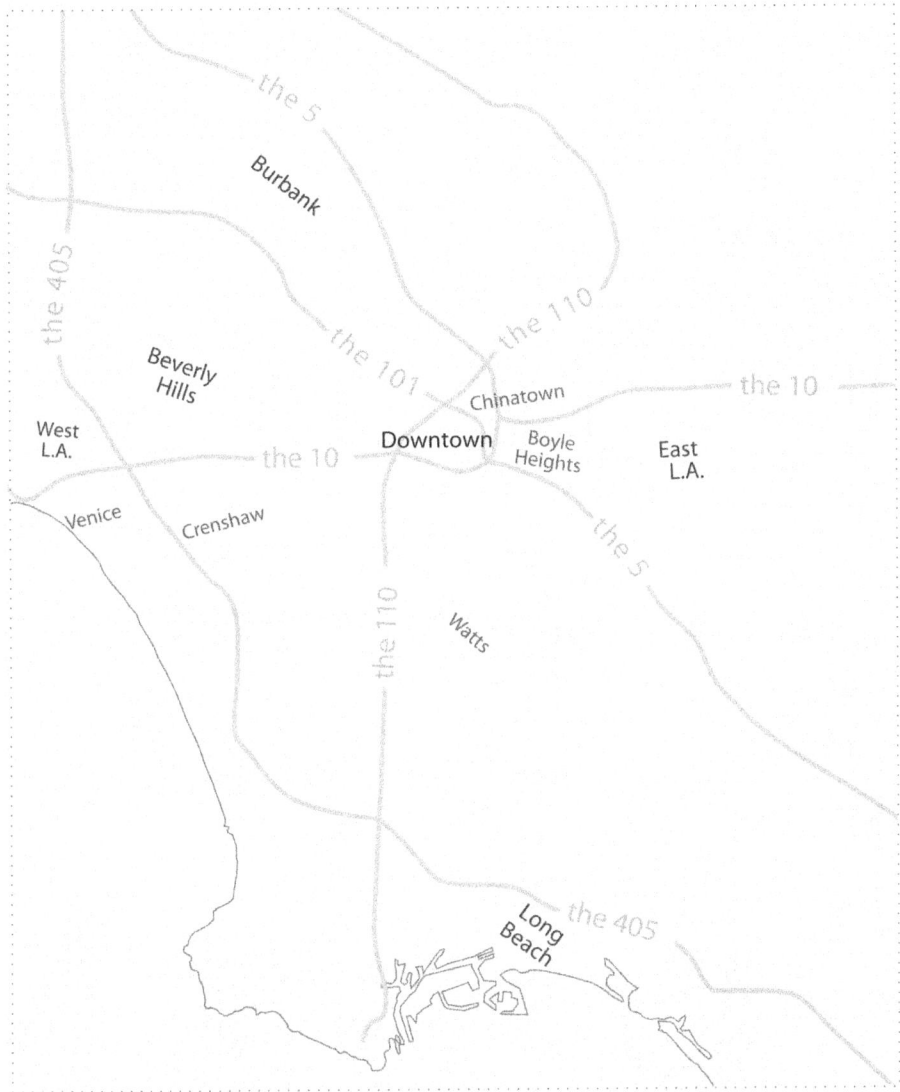

Map 1. Los Angeles neighborhoods. Courtesy of Ohio University Cartographic Center.

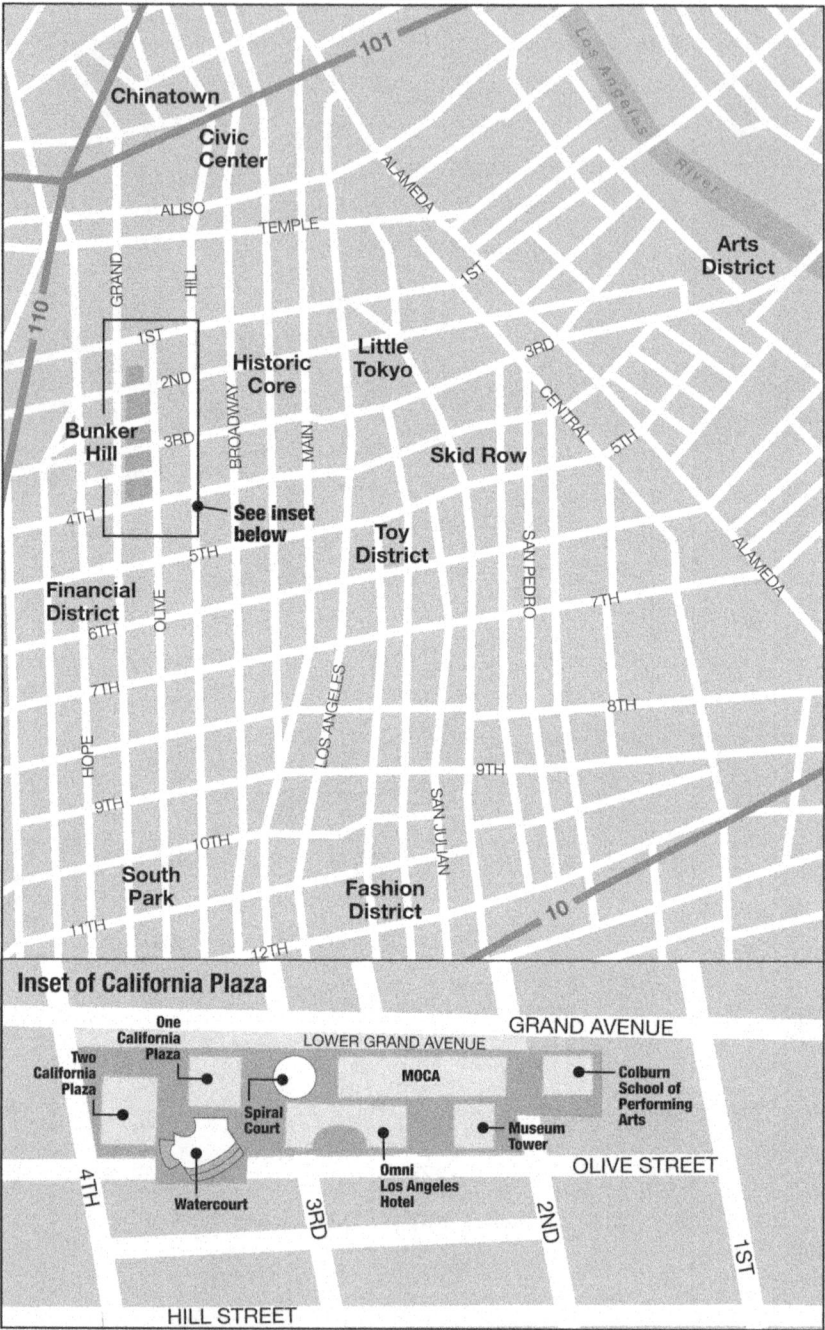

Map 2. Downtown and California Plaza. Courtesy of Ohio University Cartographic Center.

Introduction

Sounding the City

Since city making and citizen making were the same, these improvers directed their attention to citizen betterment through aesthetic reforms. Public art had long been a community educator, a stimulus to give pride and to commemorate patriotism. Harmonizing advantages would thus enable the common masses to "enjoy a constructive and vitalizing force," they would enhance and emphasize the prestige of the nation, and promote among its people the desire to excel. A happier people, a better citizen, democratically instructed and more artistic in mind and soul, would arise from this beautiful and ceremonial American city.

—M. Christine Boyer, *Dreaming the Rational City*

Looking around at the audience, she said, "This is the epitome of L.A." "Why?" I asked. "Because of its diversity," she replied.

—Grand Performances audience member

I sit at the back of the stage, the other cellists on either side and in front of me. With my right hand I pull the bow into the string in rhythmic repetition. My left hand presses the same string against the fingerboard, as I make the low sounds warm with vibrato. With the stage drained of the cascading water that runs over it when not being used for a performance, burbling sounds behind us are the only sign of the Watercourt's fountain—the centerpiece of the office towers' public space. Lit now in white, by the end of the evening colored lights will illuminate the shooting jets, turned up as the music increases in intensity. The sound of the water accompanies the viola, which we hear from our right; the violist stands, moving to the beat as he plays a lyrical jazz-inflected solo. A drum kit in the middle of the stage anchors the hip-hop orchestra. In front of us, DJs with headphones pressed between their shoulder and ear are bent over turntables. The conductor faces the orchestra from the podium, his back to the MCs who rap from the front of the stage, directing their words to an audience seated across a pool of water. Feeling the beat in our bodies, we dance in our seats, waving our bows in the air. Dusk settles over the skyscrapers of California Plaza, heightening the spaciousness and spaciness of the Los Angeles light.

The previous summer, when I was watching the daKAH Hip Hop Orchestra from the audience, it was the only performance that made the space feel transformed. California Plaza no longer had the controlled feeling of the corporate plaza, with its pink granite pavement, light green metal scaffolding, and reflecting pool with carefully landscaped planters. Young people filled the area, moving through the space and dancing, treating the space like a club. On the Upper Plaza women languorously rolled hula hoops on their hips. People passed out flyers for upcoming shows, extending the "club" space into the parking garage where they continued to hand out cards to people sitting in their cars waiting to exit. DaKAH had a driving energy level that never flagged throughout the performance: the beat was steady, the MCs stalked across the front of the stage, and the bass resonated through the plaza. After hearing daKAH I pursued the opportunity to play with the orchestra. Here was a chance to be part of this sound, this experience, to play cello in something like a big band, a ska group, an ensemble with loud, fun, danceable instruments—horns, percussion, DJs— and parts that cellists rarely get to play. Until that point I had been working with the presenters to understand the motivations and preparations for the performances. Now sitting onstage in California Plaza's Watercourt, I was performing in one of Grand Performances' public concerts.

A hip-hop orchestra, a mariachi star, El Vez the Chicano Elvis, and a Chinese modern dance company are just some of the artists in Grand Performances' free summer concert series, held in California Plaza on Bunker Hill in downtown Los Angeles.[1] Sounding a diverse city, these concerts are representations of a city that is both imagined and made in practice. Grand Performances aspires to "reflect the diversity of the city" in its audience. Its programming, though eclectic, focuses largely on world music performed by local and international performers.[2] Programming is influenced by how the presenters imagine the diversity of the city. Performances are coded by race (such as jazz), ethnicity (rock en español), nationality (Indian classical dance), or generation (hip-hop).[3] Each concert is intended to draw audiences who share the ethnic, racial, or national background of the musicians as well as others curious about or interested in that art form, those who regularly attend Grand Performances' events, and those who, as the director states, "want to be part of a multicultural group of people." The presenters hope that those who share the background of the performer will feel greater inclusion in the civic body and that others will better understand and tolerate those whose performing art form is presented.

Sound, Space, and the City is centrally concerned with considering the nature of contemporary civic life. Grand Performances provides a unique case for exploring the meanings and makings of the civic through its conditions of possibility, institutional structures, and content, as well as potentialities for membership and exclusion. What follows is the story of how such a project is created, experienced, and understood from its initial formation through its fulfillment in the experience of dancing together, an experience in which the achievement of aspirations for Grand Performances, for the city, and for contemporary public life is recognized. Woven through this trajectory is an analysis of what is involved in creating events that are understood to reflect a diverse city, as well as the work required to make these events and their conditions of possibility appear neutral. The significance of Grand Performances as a site for sounding—for imagining and creating—a real and ideal city lies in this process.

Cities around the world are undertaking similar projects as that of Grand Performances, "imagineering" neighborhoods, activities, and developments around a diverse city. Hence civic performance offers a locus for understanding intersections of broader concerns facing urban residents and scholars, including social relations and diversity, public space and civic life, privatization and suburbanization, and economic and cultural globalization. In Los Angeles, trends of privatization, downtown revitalization, immigration, and an emphasis on arts and culture that are defining features of neoliberal globalization provide a context in which international arts programming and the figuring of a local multiculturalism as international help shape Los Angeles as a global city in emergent ways.[4] The global city is now a social fact, created through material and imagined processes that range from urban development funded by international investment to the chile verde taco a downtown lawyer eats for lunch. At times taken for granted, it provides a context for projects that aspire to support global city formation.[5]

Many cities are using free concerts as a means of bringing people to newly redeveloped downtowns. Grand Performances' uniqueness lies in its civic intentions in a city perceived as lacking centrality and civic life. Making downtown a city center supports ideals of urban form and public life, against Los Angeles's infamous sprawl. The desired center is considered foremost a place where *a* public of the city can come together, a place for Angelenos to gather in order to feel like they are part of one civic body. Overlooking the Historic Core with Broadway's mixture of swap meets and historic movie theaters, loft developments and single-room occupancy hotels, California Plaza is one of several corporate projects built on Bunker Hill in the 1980s as part of efforts to remake downtown as an urban center.

Built on the razed land of Bunker Hill as part of Los Angeles's first urban renewal project, California Plaza was constructed with the intention that it would serve as public space for the people of the city. Today the Plaza boasts two gleaming high-rise office towers, an upscale hotel, the Museum of Contemporary Art, fast-food options for lunch, a coffee shop, an Italian restaurant, a series of fountains, a granite plaza, and a free summer concert series. An exemplary case of the privatization of public space and accompanying effects of security, surveillance, and exclusion, California Plaza also evinces the varied tropes of postmodern L.A., from the dominating presence of late capital reflected in its architectural surfaces to a diversity of people, buildings, and sounds (Davis 1992; Dear 2000, 2002; Dear, Schockman and Hise, 1996; Jameson 1991; Loukaitou-Sideris and Banerjee 1998; Soja 1989).

As the sky grows dark, the fountain behind us begins to shoot jets of water now colored by lights. By this time the energy onstage is high. The view from the stage is of office towers framing the sky, under which the audience, drawn from around the city, dances. The setting emphasizes the urbanism of the event, as the office towers symbolize "city" in a metropolis that is often described as not a real city because of its lack of an urban center, of pedestrian life, and of public spaces. A photo I took that night of the view from my seat shows another cellist's back graced with a large tattoo of a Chinese character in the foreground and gleaming high rises filling the background. Reflecting a diversity of style, the range of instruments in the group, and the city, the image later prompted the hip-hop orchestra's direc-tor to say, "That's daKAH right there." In performing in Grand Perform-ances' public concert series, daKAH is made as Los Angeles, makes Los Angeles a city, and makes the diverse city of Los Angeles.

Civic Performance

Later, I order a fish ceviche tostada and a lime margarita. I sit alone, alter-nately watching the wrestling on the TV above my head and looking out the front window of 7 Mares restaurant in the Lincoln Heights neighbor-hood north of downtown. My glass glistens below its salted rim, the frozen blend melting slightly. A man walks in carrying a guitar. He is slight, with reddish-blond hair, and I take him for a nerdy hipster coming in to eat in this predominantly Latino neighborhood in a restaurant where nearly all the customers are Latino. Instead he starts singing and playing his guitar. He's good. He has a presence and voice that captivates listeners, playing with a rhythmic integrity while making just the right hesitations to keep the song alive. He sings in Spanish of "amor" and "la playa." Maybe he's a white man from Mexico? A young girl eating with her grandparents is transfixed.

As I wait for the bill, he asks if I would like to hear his CD and offers me a portable CD player and headphones. "It has musicians from the Buena Vista Social Club," he said, asking if I know what that is. "I go to Cuba often," he explains. I listen to a highly orchestrated song, his voice smoother than what I had just heard, and ask if he has a CD of just voice and guitar. "No, not yet." "Are you from here?" I ask. "No. New Hamp-shire." I buy the CD for $8. "12 Canciones (12 songs), Boleros Classicos con orquesta (With orchestra)," the back advertises. There are photos of Mateo with his guitar on the streets of Havana and Los Angeles and a photo of the concrete basin of the Los Angeles River. The liner notes are musings on race and class, inequalities and injustices, tinged with despair and opti-mism. The songs address subjects such as the low wages earned by illegal immigrants and the environmental consequences of L.A.'s car culture.

This is the Los Angeles of Grand Performances: an amalgam of lan-guages and cultures, where neighborhoods dominated by a given ethnicity are frequented by others in unpredictable intersections of consumption and exchange; where Central Americans order tureens of fish soup at a Thai restaurant for lunch while five blocks away a Mexican nanny pushes white

Figure 1. View from daKAH's cello section. Photo by author.

children in a stroller, her own perched in the storage space at its rear; where some residents of West L.A. claim to have not gone east of the 405 freeway in fifteen years while the subway is standing-room-only at the end of the workday and I am the palest person in the car; where festivals and outdoor concerts present artists from the city's immigrant traditions, coordinating gatherings of people from different parts of the city and backgrounds who may or may not otherwise meet.

"Diversity" as a defining feature of Los Angeles pervades the discursive space of the city.[6] Many Angelenos will tell you stories that are variations on the one above. Diversity is used to celebrate as well as castigate, for its utopic potentials and dystopic qualities, to illuminate a wealth of contributions, and to mask exclusions (Davis 1992; Jencks 1993; Rieff 1992; Soja 1996; Valle and Torres 2000). Feature films from *Blade Runner* to *Crash* narrate a dystopic diversity characterized by alienation and fragmentation, using Los Angeles as a paradigmatic site in which to locate national concerns about the body politic in the context of globalization. Civic performance is a key space for negotiating the promise and peril of diversity. Oriented toward the people of the city, downtown civic performances aim to bring Angelenos together in the city center, helping shape a diverse Los Angeles and its urban public. At once utopic and dystopic, emergent spaces of belonging are created through inclusion and exclusion.

In a city in which free performances can be heard any summer evening from the Santa Monica Pier to the Hollywood Hills, from Culver City to San Pedro, Grand Performances is the only public concert series intended for the people of Los Angeles.[7] An underexamined public art form without, as yet, a dedicated body of literature, in their ideal form public concerts are free, outdoor concerts in a public space in the center of town, with music intended for a general public, supported by the municipal government. These defining characteristics of public concerts provide the conditions of possibility for Grand Performances to orient its concerts for the city, to serve as a civic institution, and to be recognized as such. "Civic performance" thus describes free public performances that are intended for the city and that aim to improve the city and its citizens through performance. Civic performance also captures the multiple ways in which the civic is performed at California Plaza, by musicians and dancers onstage, the work of the presenters, and by audience participation. As a nonprofit in a corporate plaza, Grand Performances stands as a civic proxy under neoliberalism.

Historical cases provide a model for understanding what makes Grand

Performances a civic institution, indicating continuities in a nexus of culture, state, and market as public performances provide a space in which a unified civic body is imagined and enacted (Boyer 1983; Ryan 1997). These cases suggest the continuing significance of public concerts for understanding what a given moment's social ideal is, and how that ideal conceals as well as reveals inequalities, exclusions, and critiques. At the same time, framing current public concerts as one moment in a linear history of a form risks flattening the issues raised by civic performance at various historical moments, which, while similar, are also unique to particular times and places.

In the European classical tradition, public concerts referred to performances that, beginning in the seventeenth century, were open to a general public for a small fee rather than only for the court or another limited audience (Johnson 1995; Notley 1997). In the United States, an older civic form, like today's public concerts, helped create and newly define a "public" that should have access to things like music and "common" spaces (Crouch 1981:8).[8] In shaping urban publics around geographies of centrality and marginalization, Grand Performances sustains elements of an earlier American tradition of public concerts. At the end of the nineteenth and beginning of the twentieth centuries bandstands were built in town squares as performance venues for the brass band movement that swept the country (Starr 1987). While brass bands were a form of popular entertainment, by the Progressive Era, in downtown Chicago free summer concerts presented a mixture of accessible and challenging classical music that was intended to both entertain and uplift its listeners. Helping spatialize ethnicity in the city by sounding an ethnic urban geography, music from the country of origin of local residents was presented in neighborhood parks (Vaillant 2003).

In recuperating past ideals for present purposes, public concerts acquire symbolic, representational meanings. Such efforts often come packaged in sets of ideals harkening to a past that might never have existed. This lost public is a "phantom public" (Robbins 1993), a "haunting" (Derrida 1994) that provides motivation for emergent public practices and institutions in the contemporary moment. Just as urban theorists and planners like Mumford and Le Corbusier drew on a model of Athens that took the agora as "a symbol of urban society in general" (Lefebvre 1996:98), notions of democracy and public life associated with other times and places are manifest in the public space of the corporate plaza. Hence intentions of openness associated with earlier public concerts are recuperated as part of a contemporary project of urban public making, in which ideals of public life are constituted and supported by free performing arts. Though today's concerts draw on older models to assert their civicness, their projects are specific to a contemporary context, which, nevertheless, is not totalizing. With late twentieth-century concerts, multiculturalism and civic culture work along-side, are part of, and help support—though are not reducible to—a neolib-eral project.

As participants in civic events, public concert audience members are civic subjects. The term "civic subject" marks a critical terrain of subject formation related to citizenship that is also fundamentally of the city. With citizenship understood as the relationship of subjects to the state, "civic" situates this membership in the city. The city has taken on a privileged role as a site for the production and negotiation of the terms of national citizen-ship. Though formal citizenship resides with the nation, cities can be "the sites of the most profound questions of belonging and identity" (Rogers 2000:287; see also Holston and Appadurai 1996; Isin 1992, 2000; Sassen 2003; Taylor 1994).

In cities, difference, conviviality, and aggression converge to create spaces of formal and informal membership, and civic subjects are figured in multiple sites that are not necessarily explicitly political. Residents with-out legal status bring their children to public schools, formal citizens pay cash for landscaping, and everyone is a potential public concert audience member. Urban public space is a key site for the expression and negotiation of citizenship in cities even as it reflects a gap between urban residents as formal citizens and an urban public that does not have a legal relationship to the state. Similarly, though ideal types of public concerts might be run by municipal governments, they are not sites of formal political participa-tion. When public concerts are not run by the city, even insofar as their audience is "the city," an argument for a connection to formal citizenship becomes more tenuous. The category "civic subject" captures the civic potential of public space and performance, where membership is organized and authorized through modes and forms of recognition, use of public services, and participation in public events.[9]

Public concerts encourage civic participation based on consensus (Mar-tin 2004; Schmidt 2005). The etymological roots of "concert" describe par-ticipation and consent, social gathering, and harmonious interaction. Since the mid-nineteenth century, a concert has entailed a formalized, reproduc-ible, and shared social practice around a performing art form. While by definition a concert "makes music the centre of social attention" (W. Weber 2001:221), a dance performance can also be called a concert. Following, while music defines the events and Grand Performances as an institution, its season also always includes dance and spoken-word performances. Shifting its assessment of its programming, in the beginning of the 2004 season Grand Performances changed its self-definition on its Web site from "public concert presenter" to "free performing arts presenter."

At Grand Performances' events critique is limited as only the fulfillment of ideals is recognized: people see heterogeneity and diversity in the audience but not the exclusions out of which the public is wrought, they feel part of a collective and do not see the corporate support, and they perceive the space as public, open, and accessible to all. Moreover, what is not seen is how the formation of urban public life around consensus entails the exclusion of dissent, just as the inclusion of newly recognized subjects and publics entails the exclusion of other categories and bodies that do not conform to these standards.

A multicultural public is in principle inclusive and open to difference. Yet as *the* urban public it occludes both the exclusions on which it is founded and the histories of its making. At Los Angeles's downtown public concerts, multiculturalism's emphasis on ethnic and racial diversity excludes class as a category of diversity and excludes people who are marked by class (Michaels 2006). Explicit expression of a political position is considered divisive and is excluded other than under specific circumstances. At stake for this study are the bases of exclusion and how those exclusions are occluded—how, that is, a multicultural urban public is produced as normative or general, and how this process is observable in and supported by Grand Performances.

Sounding

"Sounding" captures the ways in which Grand Performances' articulation of a diverse, global Los Angeles and its urban public is a *spatial* and *sonic* process. At civic performances, music, urban space, and subjects converge and emerge, as the city comes into being through interweaving sonic and spatial dynamics. Space, Lefebvre writes, is "first of all *heard*" and "*enacted*" before it is seen (1991:200). With his project on rhythmanalysis, Lefebvre evocatively draws from musical language to explore overlapping temporalities of urban space (2004). Rhythm, unfolding yet not necessarily regular or repetitive, marks tempi of motion and histories of circulation, revealing the temporality of space and the spatiality of time. Visible and invisible rhythms of movement and flows converge: looking out his Paris window, Lefebvre observes the changing stoplights, people walking across intersections, cars speeding around traffic circles, and banks through which multinational capital invisibly moves (Lefebvre 2004:27–37). Layers of history are sedimented in the built environment, revealed by monuments or concealed for new purposes.

At downtown public concerts, histories and ideals coalesce, as multiscalar spatial and temporal rhythms, taking more and less apparent forms, are seen, heard, and felt. Multiple social and musical rhythms converge to create the space of California Plaza through practice. Janitors sweep the granite floor while gardeners maintain the landscaped islands in the fountain, letting the plants go unfertilized on the island chosen by a family of mallard ducks to hatch their eggs. Children run squealing through a fountain that shoots jets from the ground in an unpredictable pattern. A businessman takes a noontime nap on a riser of the Marina Pavilion.

Dynamics of musical meaning inscribe processes of city- and civic subject-making.[10] Each person who comes to a performance brings personal and social experiences, identifications with music and the city, and ways of being in the city and in the world. Performances realize long histories of practice and expression that are deeply rooted in social formations, organizing belonging along lines of subject, culture, and nation. Heard and felt in the body, an aural city is experienced through noise and music. Sounding the city, performances reach out to draw people from around a diverse yet segregated city to downtown concerts. Taking place in time and over time, sound moves people spatially and affectively. Sounding out the sentiments of a crowd, civic performance moves people to dance, an experience that becomes a locus for personal and civic identifications.

Embedded in the built space of California Plaza is a planning history that involved making representations of space and enacting them in the built environment while imagining ideal social formations that might be realized through public performance. A project to remake downtown as a center for Los Angeles took shape through urban renewal and redevelopment. Urban renewal sounded city space by clearing the land and dispersing Bunker Hill's residents to peripheral neighborhoods. In helping shape a segregated city with a downtown posited as "neutral," this project laid a foundation for Grand Performances. Traces of this history remain,

inscribed in the built environment and articulated in aspirations for attracting a public of Los Angeles to performances. As discussed in Chapter 1, the
purported neutrality of downtown plays a dual role of organizing urban
space and of supporting diversity as a normative and defining feature of
the city. Ultimately the development of California Plaza contributed to the
formation of *an* urban public in a city defined by diversity.

The center-making project reflected in Grand Performances is geographic, social, and conceptual, entailing the construction of Bunker Hill
as an urban center, an emergent perception of the city's public as defined
by diversity, and the naturalization of both as normative. Now diversity as
a defining feature of Los Angeles is supported by an urban geography organized as segregated, with ethnicized neighborhoods surrounding a center.
This neutral center in turn supports the coalescence of a diverse group
which, drawn from those neighborhoods, might exemplify the imagined
diverse city. The removal of unruly bodies from Bunker Hill during urban
renewal is sustained in the privatized public space of California Plaza, where
rules regulate the bodily comportment of members of the public and security guards encourage the homeless to move on.

A technique for navigation and communication in visually impenetra-
ble ocean depths, sounding is at once sonic and spatial, using sonar to
delineate a known and bounded geographic area out of an otherwise
unknown expanse. Chapter 2 focuses on how free downtown concerts reach
out from the center to demarcate Los Angeles's neighborhoods in order to
bring residents otherwise separated by neighborhood, identity, and music
together at performances. The city is made through embodied practice;
once relocated from the future site of California Plaza, Angelenos now
move through the city to attend free concerts of artists from around the
world. By tapping into media that imbue the lived and the everyday, Grand
Performances articulates the rhythms of a city even as it presumes a fixity
of correlations between geography and social groups, media and identity
that is supported by mental and printed maps of a segregated city. The
international nature of Los Angeles's multiculturalism is articulated as marketing reaches out through radio waves and print of ethnic media, sounding
geographies of media publics and ethnic subjects brought together as public
concert audience members. Inscribing a global city, marketing and media
intersect with immigration, language, and translocal politics.

Ultrasound focuses the technology of sounding on the body, making
visible that which is otherwise unseen. Though the physics of music is the

movement of atoms in space, it is never possible without a medium, which, whether a musician, a CD, a radio, or an instrument, is always social. Brought into being through practice, sound supports the making and unmaking of identifications of musical meaning and urban diversity. Chapter 3 focuses on a performance by the daKAH Hip Hop Orchestra in which I played cello to discuss how the group's diversity, framed as social and musical, is expressed through composition, musical styles of its members, and DJ lyrics. In this context learning to make the cello sound like a drum machine brought together hip-hop and classical genres through the embodied practice of the performers, creating a musical space for social transformation across generations.

Presenters, artists, and audiences largely concur that performance *means* something. In particular, there is agreement that performance—of music or dance—is bound to social identities organized around common national or racial origins, and that those identifications can be read within the context of Los Angeles. Hence musical diversity allows for parallels to be drawn between performance and the city, invoking and creating a diverse Los Angeles. DaKAH also articulates another global city, as inequality, racialized urban geographies, and diversity—all in part effects of postindustrial capital—are invoked by a now global popular culture form with a deep history in Los Angeles's neighborhoods and music industry. Here global capital and multiculturalism are two ends of a spectrum of globalization that imbues and draws from urban, national, and transnational dimensions of racial politics inscribed by urban segregation and immigration, and of diasporic spaces that link cities, ideologies of civic space, and city-making from Beijing to Boston, Lagos to Los Angeles.

Sound shapes subjects who, in listening, resonate as sonorous selves (Nancy 2007). Chapter 4 explores the intersections and divergences of performance and politics that figure public concerts as civic spaces and audience members as civic subjects. Representation, recognition, and participation, tenets of democracy, also structure the ways in which civic performance reflects a diverse city. Identifying the performance of what some call "my music" shapes civic performances as sites of belonging. Extended to others, "my music" further supports public concerts as civic spaces organized around agreement, harmony, and tolerance. Correlations between music and identity are performed through modes of identification and interpretation. Inscribing a relationship between the two supports the formation of a moral community of an urban public and an imagined

Figure 2. California Plaza during a performance. Photo by Guilherme Rafols.

diverse city. Neither mimetic nor epiphenomenal of a separate political sphere, public concerts are spaces for the negotiation of social and political concerns, for terms of political identification and engagement.[11] The civic of Grand Performances is a space of consensus that excludes the noise of dissent, whether in the form of political debate or challenges to the terms of the project. Hence electoral politics are excluded other than in select cases, performances are oriented toward the taste of a general audience, and a general agreement is maintained that performance has social meaning.

Weather balloons carry instruments into the atmosphere that take a sounding to determine conditions that are translated into meteorological information. At Grand Performances, volume, measurable in decibels, is increased to encourage people to dance as well as deployed for noise ordinances that structure urban civilities. Affective identifications with musical genres attract people to the concerts, where an experience of dancing with strangers facilitates the recognition of a diverse and global Los Angeles in the audience. The public concert audience is constituted as a collective, affective musical public and a civic moral community through embodied experiences of listening, dancing, and feeling moved by the music. Chapter 5 focuses on this moment, considering how ritual theory is invoked as a means of articulating contemporary aspirations for community, togetherness, and diversity. At the same time, this chapter aims to theorize the audience of a live event as a collective, as meaningful, and as a locus of meaning making. Aspirations for a utopic diverse city become real as they are experienced.[12] Outside of everyday life but not strictly opposed to it, the public concert provides a space in which an ideal of and for social relations, and for the city, is realized through experience. Hence Grand Performances' events create a space in which an imagined diverse Los Angeles is manifest in the experience of audience members, who participate in and help create a musical space that expresses that ideal for a time. Today Angelenos come to free summer concerts in California Plaza on Friday and Saturday nights, enjoying the strains of the shakuhachi, cello, or saxophone as part of a multicultural audience that some identify as "the epitome of L.A."

1

A Center for a Centrifugal City

Merely vacant land now, Bunker Hill can begin to tie the
disparate strands of government center, cultural center,
popular markets, ethnic communities and burgeoning
business core together—so that it becomes the center for
all these other centers. . . . Only by means of a project of
such scale and future mindedness can Los Angeles express
what is unique about itself and at the same time begin to
fulfill its future role as a center of centers of the western
world.

—Bunker Hill Associates, 1980

Every city needs a place that the whole community can
call its own: a space where people gather for celebrations,
where people gather to relax and perhaps where people
gather for free concerts and theatrical performances.

—Grand Performances, 1997

Looking out at Los Angeles at night from the Griffith Park Observatory,
densely sparkling lights span the expanse of land as far as one can see to
the east, south, and west. North, the mountains are sparsely lit, the lights
of the San Gabriel Valley just visible on the other side. Visitors point out
landmarks, places they know, inhabit, experience—where they live, the
ocean, Hollywood. Tonight the air is heavy and hot. One has the feeling of
being inside a dry oven. Downtown is lost in the expanse of lights, unless
you look for it. No one near me mentions it, either as home or a place to

go. The lights of downtown's skyscrapers—Library Tower, California Plaza, the Aon Building—rise vertically from the flat surface, evident as a central business district but appearing as one small part of this wide expanse.

This bird's-eye view does not convey the flurry of past and present activity downtown, the rhythms of urban life that have coalesced in the center city. The intensity of attachment to a project of making downtown a center for the city. The sedimentation of a century's history of growth and decline and regrowth visible as signs advertising loft developments grace façades that have remained the same since the 1950s. Mid-century demolition of Victorian mansions already turned boardinghouses. Traces of global flows of capital, media, goods, and people materialized in stacks of hundreds of plastic sandals and offices of multinational corporations, in the sound of Spanish as the lingua franca of street commerce and the largest homeless population in the country. A skyline without skyscrapers greeting a future developer and real estate mogul as he flew into Los Angeles in the late 1960s. Bison parading down Olive Street as camera crews follow, capturing the sight for a beer commercial. The seriousness and intensity with which developers, neighborhood councils, residents, and property owners imbue their discussions of the future of downtown. Ongoing struggles over space by homeless, artists, politicians, and business owners, as the route of the monthly art walk increasingly overlaps with the boundaries of Skid Row. An audience of Angelenos dancing to the strains of raï or rock en español or Afro-pop at California Plaza's Watercourt, as the color of the sky deepens to the blue-black of evening and the lights illuminating the water behind the performers change from yellow to purple to red.

At once symbol, site, and structure, the center is mythical, occluding the past and present processes of its formation (Barthes 1972; Bourdieu 1984). The center is constituted as such in its definition and appearance as neutral, thereby obscuring its own making and the exclusions wrought therein. Today, a board member and local developer maintains, Grand Performances "is a complete success in the development saga of Cal Plaza. Standing at the end is this small piece that's fabulous free music. It's like a blade of grass after Mount St. Helens." This blade of grass provides a metaphor for the ways in which the current success of Grand Performances depends on naturalizing the result of a long process that has included urban renewal, bankruptcies, and failed aspirations; with the blade of grass the focus, the violent eruption still smoldering beneath the surface is concealed.

Yet the center has forms through which center-making processes can be investigated. Grand Performances' achievement of being "recognized as outstanding in its efforts to promote the performing arts to the public and establish ties to the diverse communities within the urban area" (CRA/LA 2000:2) emerged from a longer history of crafting downtown Los Angeles's Bunker Hill as a center for the city. Though Los Angeles is stereotyped as lacking a geographic center, this does not prevent centralizing efforts that privilege an alternate discourse of urban form. Making downtown an urban center helps construct the city in a model made famous by the Chicago School, in which the city radiates out from a centralized core (Park and Burgess 1925). This center structures a normative city and its public through intersecting material, social, and imagined processes.

The Bunker Hill Urban Renewal Project was part of a larger endeavor to re-create downtown Los Angeles as a center that would be used as such. The terms of *re*newal, *re*vitalization, and *re*development that define the project imply a past to which the population displacement, land clearance, and high-rise construction was returning. Yet the project was always one of improvement and progress, of providing the city with the center it supposedly needs, and of putting populations in presumably proper places. Once there, residents are available to be called upon as public concert audience members and to be reshaped as civic subjects.

Legitimized and authorized presence in the center structures membership in the urban public, which comprises those who have ownership of the city, represent the city, and maintain the right to the city (Lefebvre 1996). The public space of California Plaza was conceived of as a place where the public of the city would gather; negotiations over the built environment and programming content helped define the public of the space and of the city. The plan that was ultimately implemented as Grand Performances focused on presenting performers who would draw a multicultural audience, helping define the public of Los Angeles as diverse. Today the privatized public space of California Plaza provides a context for shaping bodies in and as the urban public through relative inclusions and exclusions and the disciplining of disposition and comportment.

Conceiving of downtown as "neutral" makes it possible for a corporate plaza to serve as civic space, allowing the concerts to attract audiences who can be recognized as reflecting and representing the people of the city as a whole. Downtown, the presenters assert, is the only place where a diverse group of people can gather without any single group marking the area with a specific ethnicity. Because it is "neutral," it provides a place where "all populations feel welcome and feel a sense of ownership in their city and their space." Moreover, they say, "This neutral space can be created, established, and maintained through the incorporation of the performing arts." Urban renewal and concert programming are thus two sides of a process of cleansing urban space to create neutrality of place and public, the former an act of physical land clearance, the latter a process of public formation that excludes through planned inclusions. Both sound the city, presuming and producing a diverse Los Angeles that is constituted by an initial centrifugal spread of spaces and subjects brought back together downtown for free summer concerts.

Constructing Neutrality

Overlooking the old downtown where vacant banks, empty department stores, and unused movie theaters lined the streets of what is now designated the Historic Core, Bunker Hill provided the possibility for a new downtown whose use and populations might be structured through the built environment and public events.[1] The Bunker Hill Urban Renewal Project cleared the area of undesirable people, providing a seemingly blank slate for a new center for the city. Urban renewal civilizes the city through a people-moving project, creating an uneven geography that values land over residents and designates the worth of people according to their location in the city. In downtown Los Angeles, the "wrong" people were moved from the city center in the renewal process and the "right" people brought in as part of revitalization and redevelopment. By dispersing low-income, minority residents to other parts of the city, it contributed to the formation of a diverse yet segregated city.

Bunker Hill had been a site of potential and realized reform and planning projects since the 1930s that focused alternately on the eradication of slums and property acquisition for public use. The first swing of the wrecking ball in 1962 marked the culmination of these efforts and the realization of the Bunker Hill Urban Renewal Project, drafted in 1958 by the Community Redevelopment Agency/Los Angeles (CRA/LA). The CRA/LA's mandate has been, and remains, to recognize and eliminate "blight." Though the term was used in discussions of urban policy as early as 1930 (Gordon 2004:310), the United States Housing Act of 1949 facilitated its transformation from idea to practice by providing funds for interventions into the urban environment on the basis of blight (Fogelson 2001:376). The Bunker Hill project was the first undertaken by the CRA/LA under its mandate.

The vagueness of the definition of blight supports its effectiveness as normative description and prescription. "Blight," the CRA/LA maintains, is "deterioration of an area caused by physical, economic and social forces" (CRA/LA 2005). An "urban disease," blight reflects a confluence of social and spatial phenomena ranging from juvenile delinquency and overcrowding to winding streets and unmaintained homes (M. Scott 1949:108). Deployed as part of Progressive Era reform efforts to ameliorate slum conditions, blight was taken up in the service of the interests of downtown stakeholders whose businesses were suffering as part of a national pattern of suburbanization. A multifaceted definition of blight emerged to meet the interests of downtown business owners, municipal governments, and public housing advocates (Fogelson 2001:317–80). By the time it was implemented as part of urban policy, blight referred not only to poor living conditions but also to "inappropriate" use and financial liability for the city. It is perceptible in dilapidated buildings and conditions that reduce revenue or hamper economic development.

Occupied by Victorian mansions, Bunker Hill in the beginning of the twentieth century was home to Los Angeles's elite. As the city grew the wealthy moved westward. By 1950, according to the Community Redevelopment Agency/Los Angeles, "Bunker Hill was an aging, hilly pocket of downtown, filled with deteriorating rooming houses, residential hotels, and corner stores" where "only 18.4 percent of the dwelling units were considered 'acceptable' or livable" (1990:5; see also Adler 1964; Davis 1992; Fogelson 2001; Hylen 1976; Loukaitou-Sideris and Sansbury 1996; Pugsley 1977; and Wild 2005). Overall, Bunker Hill neatly exemplified a contemporaneous description of a blighted area: "At an accelerated pace the whole area goes from bad to worse, until it becomes known to city officials and social workers as a problem area, in which local government spends more to curb crime, fight fires, treat disease, and provide financial assistance to the needy than it collects in taxes" (M. Scott 1949:110).

Notions of social reform were wrapped in a vision of modern urban rationality: clean houses would make clean people, in both body and morals. The concept of blight assumes a material determinism that insists the habits and lives of residents will improve if negative features of the built environment are eradicated. Eliminating crowded housing and unplanned neighborhoods supports a normative model of urban sociality that promises improvement of bodies, houses, neighborhoods, and city. Rationalized modern planning on the level of the street, the neighborhood, and the city is intended to contribute to mental health, better lifestyles, and democracy. Blight figures prominently in *Metropolitan Los Angeles: One Community* (M. Scott 1949), which draws a plan for the city out of existing features and articulates problems as the converse of norms. Suburban ideals of single-family homes with large lots are provided as a solution for overcrowding, defined by bedrooms shared by more than two people, and the correlation of blighted areas with areas of high population density. Maps graced with black dots showing that both blight and population density are highest in and around downtown support the development of an urban form in which downtown is intended for business and sprawling surrounding areas for residential use (M. Scott 1949:48, 109).

A rhetoric of care, ideals of modern living, and the relocation of the poor from potentially profitable urban spaces to segregated neighborhoods went together. According to the CRA/LA, redevelopment promised much: the residents of Bunker Hill would no longer have to live in the blighted conditions that caused disease, crime, and juvenile delinquency. They would be relocated to a "safer and more healthful residential environment" (1958:25). And they would be provided with new, modern homes that would ostensibly be as affordable as those they currently occupied. Desire for a clean and profitable city facilitated the spatialization of a racialized Los Angeles. By dispersing its affordable housing projects, the CRA/LA continues to shape a city segregated by race and class that is key for making downtown "neutral." A Community Redevelopment area is funded by tax increments, and the CRA/LA accrues any difference between taxes due on the properties before and after redevelopment to fund its projects in that and other redevelopment areas in the city. California's Community Redevelopment Law requires that at least 50 percent of the revenues must be used for affordable housing, presumably, although not explicitly stated, for those displaced by the redevelopment project.

Bunker Hill funds for affordable housing have primarily been used to fund single-room occupancy (SRO) housing on Skid Row and housing projects away from downtown, especially in the low-income neighborhoods of Boyle Heights, South Los Angeles, Westlake, and Pico Union, thereby helping create a dispersed and ghettoized ethnic geography of the city. New development of market-rate housing on the remaining undeveloped parcels of Bunker Hill—for which a development team was selected in 2004—

continues this trend of dispersal, providing funds for "the production of affordable housing in the Ninth Council District south of the Santa Monica Freeway" (CRA/LA 2000:18); in other words, in an already poor area a few miles away from Bunker Hill and on the other side of the border created by the freeway that separates South Central from downtown.

Angelus Plaza, low-income senior citizen housing, was the only affordable housing built on Bunker Hill as part of the redevelopment project. According to the CRA/LA, Angelus Plaza "furthers the removal of blight by implementing the goals and objectives of the Bunker Hill Urban Renewal Plan that call for replacing substandard housing with decent, safe and sanitary housing for all income and age groups" (2000:10). Senior housing is a popular way of fulfilling low-income housing requirements. With diversity extended to include age, senior citizens safely meet the demand for diverse residents without bringing the imagined possibility of undesirable or criminal activity.

A project organized around blight results in the rationalization and reorganization of urban geography through the destruction of the built environment at the expense of people whose lives are improved in name. Supporting the interests of a general public over those of existing residents, community redevelopment agencies are granted legal rights to acquire property by means other than or in addition to a market purchase. While effort is ostensibly made to pay fair market value, the CRA/LA can use eminent domain to acquire property by condemning it (1958:12). On Bunker Hill the CRA/LA promised that houses deemed "worth salvaging" would be relocated, though without their inhabitants.[2]

The Bunker Hill Urban Renewal Project helped replace particular people with an abstract general public, positing the needs of the latter over those of the former. Concern for city revenue frames the difference between existing residents and an urban public. Razing the buildings in the area and redeveloping would convert "a tax liability to a tax asset for the people of the City" through the "creation of a plan of land use of great benefit to the people of the entire Los Angeles metropolitan area" (CRA/LA 1958:25–26). A notion of a general public of "the people of the entire Los Angeles metropolitan area" serves in part to mask the motivations of the city as a growth machine, of a capitalist city that values planning over people (Jonas and Wilson 1999; Logan and Molotch 1987; M. Smith 1988; R. Weber 2002). A future, abstract public is one with neither content nor political representation. Demolishing the buildings on Bunker Hill in the name of this public is clearly in the interests of those condemning the land rather than those living there; the former make a normative urban public in their image by speaking for a general public that cannot speak for itself because it does not exist as such, over the claims of those who can speak but whose voices and interests are not recognized or authorized.

Founded in 1948 by the City Council of the City of Los Angeles under the authorization of the California Redevelopment Act, the CRA/LA works in the name of the public. Even if a project is profit driven, it is framed as a public good insofar as the profit is for "the people."[3] And as work is shifted to the private, it is required to include public elements. In this way the CRA/LA integrates interests of capital into the public and figures the public in the service of the private. More than a carving out of the public, this work shows how public and private are never actually opposed but are always inextricably linked. Based in a notion of public and private as an opposition between state and market, public good and profit, the public is created as it is named, invoked as it is evoked. As "ideal types" of public and private are seemingly unmade in practice, their defining features are reinvoked, helping shape them anew.

California Plaza is now private property on public land, with a ninety-nine-year lease from the CRA/LA. Defining elements of publicness support this privatized public: urban renewal was undertaken in the name of increased tax revenue for the public while nonprofits are considered more open than government agencies. Restrictions generated by privatization facilitate public formation through relative inclusions and exclusions, from urban renewal to privatized security and surveillance. Already posited in name, the particularity of the "general public" was negotiated and clarified through a series of planning and development documents addressing the nature and content of California Plaza's privatized public space and event programming.

A Place to Gather

Development proposals read as reality writ as fantasy. Their future tense asserts an imminent reality of present possibilities. The plans for what would become California Plaza deploy representations of the built environment in the service of an imagined lived space of social activity. They posit neutrality by presenting abstract space as a background against which a general category of "people" can play. In one, planters grace dividing walls, and a businessman and businesswoman meet for lunch, leaning over an umbrellaed table to talk. Public space and events focus the fantasy of a vibrant space bustling with crowds of people enjoying themselves. These spaces convey an urban ideology of cities as spaces in which people gather and encounter each other.

A public space for the people of the city with events that would draw them there served a larger purpose of city- and civic subject-making. As imagined by the CRA/LA and developers, the public space was for the people of the city, who, in gathering therein, would make downtown a city center and Los Angeles recognizable as a world-class city. Publicness of place and people was asserted within the public-private partnership in order to achieve these goals. In 1979 the CRA/LA issued a request for proposals (RFP) for the five-block area on the southeast corner of Bunker Hill that would become California Plaza (CRA/LA 1979). Stressing the "public" part of the Plaza, the CRA/LA defined the size and scope of the area and demanded its publicness be apparent in the built form: "Central to [the 5.5 acres of open space in California Plaza] is to be the Central Performance Plaza serving as the major open space" for the development (CRA/LA 1981:3). The Central Performance Plaza should "provide a public forum, offering a natural stage for civic festivals, concerts, dance, theatre and televised shows. It should also provide a setting for informal types of activities, such as concerts and outdoor exhibitions" (1981:9). If these directives were followed, aspirations might be realized of making California Plaza a center for the city, the region, and the world.

Negotiations between the CRA/LA and the developer over the nature of the development reflect a dual process of facilitating the integration of the private into the public while maintaining the public in the context of privatization. Though public space was mandated in the request for proposals (RFP), the emergence of its form and desired users reflects the continual making of meanings of public and private and of the nature and content of Los Angeles's urban public. A series of proposals defining and refining event programming helped shape an urban public through the performing arts, by the end of which a clear emphasis on attracting a multicultural audience to Bunker Hill emerged.

A year after the RFP was issued, eighteen years after Bunker Hill demolition began, and the year Los Angeles surpassed Chicago to become the second largest city in the United States, Bunker Hill Associates won the competition to develop what is now called California Plaza.[4] The RFP had stated that the area was to "become the focal point of the Central Business District and metropolitan Los Angeles" (CRA/LA 1979:1). Both the winning and losing plans were perceived as meeting the aim of creating a new "center" for Los Angeles, and rhetoric to support that end was used to frame and evaluate both.

Bunker Hill Associates proposed a mixed-use development that satisfied the requirements of the RFP: residential towers, a hotel, office towers, retail establishments and restaurants, Los Angeles's new Museum of Contemporary Art, and a central public space. Arthur Erickson, architect for the winning proposal, proclaimed that California Plaza's potential as a center of the city could make Los Angeles "a center of centers of the Western world" (Bunker Hill Associates 1980:1). An architecture critic concurred that the winning proposal's modernist approach of "creating large open spaces by placing towers in a park, with a linear slab building reinforcing one edge" (Goldstein 1980:46), would draw the public there and thus situate California Plaza as a center. A place for the public to gather lent a social quality to the city that she otherwise found lacking: "Los Angeles is a city where there is, literally, no real *place*. There is no real gathering place in Los Angeles—no place to celebrate New Year's Eve or the Fourth of July. The most encouraging aspect of Bunker Hill Associates' scheme is that it tries to create a *place*" (Goldstein 1980:46). The winning proposal, with its uniform design, aspired to tie together disparate aspects of Los Angeles with a rational amelioration of the area's former "blight." Its modernist approach was sustained by its "uniqueness" and "future mindedness." The plaza would allow a crowd to form that would be visible as such: measurable, controllable, and identifiable as the public of the city.

Maguire-Thomas's "varied" plan, supported in a CRA/LA Minority Report, emphasized the experience of the individual participating in private, out-of-sight activities. Postmodern architecture implemented in a corporate plaza, the plan incorporated the work of multiple architects and placed modern sculptures prominently throughout the area. A "people oriented, exciting, dramatic, playful and varied urban design," it provided "the greatest opportunity to create *the* Center of the City of Los Angeles" (Bunker Hill Task Force 1980:3). The plans' differences were articulated in language that described on the one hand a gathering place for major holiday festivities and on the other a "people oriented" space, a place for a mass public and a place for social interactions between individuals. The former was valorized for its seemingly clear yet grandiose potential to both represent and create a center for the city, made tangible in the built environment and the presence of the public.

Arts supported aspirations for California Plaza's centrality. As " 'material' shapers of cities" (Whitt 1987:16), arts structure urban development and neighborhood characteristics, capital flows and the movement of people. The use of the arts in Los Angeles's downtown development reflects a strategy found in other cities across the United States and, increasingly, around the world. Though the arts have a much longer history as part of urban public life (D'Accone 1997; Johnson 1995; Strohm 1985), in the late twentieth century they became an integral part of the urban growth machine, used to draw people and attention on urban, regional, national, and international scales (Bianchini 1990; Harvey 2001; Kotler 2002; Ward 1998; Whitt 1987; Zukin 1982).

The RFP, calling for "creative integration of residential, commercial and office type uses with artistic, cultural and recreational uses" (CRA/LA 1979:29), explicitly required the inclusion of the Museum of Contemporary Art (referred to then as the Museum of Modern Art). Though the RFP did not specify that there should be a public performance space, sources inside the CRA/LA made it known privately to the developers that the inclusion of such a space would make a stronger proposal. To this end, winning designs for California Plaza initially included three major arts components: the Museum of Contemporary Art, a resident modern dance company, and a public performance series. Each of these projects was implemented through a public-private partnership, situating the arts in a shifting dynamic of public and private that shaped the value and meaning of the organizations and their respective publics. Though not all survived, their perduring presence in the built environment of California Plaza conveys histories of multiple modes of center-making through the arts.

A consortium of Angeleno power brokers that included Mayor Tom Bradley, council member Joel Wachs, and Marcia Simon Weisman spearheaded a mobilization to build the Museum of Contemporary Art (MOCA), deemed important for making Los Angeles visible nationally and internationally as an arts city (Berelowitz 1991). The museum was intended to be compatible with existing and future elite arts institutions on Bunker Hill, including the Music Center and Disney Hall. The board selected Japanese architect Arata Isozaki to design the museum's permanent home in the northern part of California Plaza. Conflict emerged around questions of whether and how much Los Angeles artists would benefit from the museum. A board of local artists, chaired by Robert Irwin, contributed to early phases of the planning process, including vetting the design submissions for California Plaza. The artists' preference for the Maguire-Thomas plan did not sway the CRA/LA in its decision. The committee was also not part of decisions about the architecture of the museum building. However, they were instrumental in identifying a space for the "Contemporary Temporary" in Little Tokyo. Initially used for MOCA shows before the main building was completed, the space remains an adjunct to the Grand Avenue museum under the name Geffen Contemporary.

Bella Lewitzky's company, the premier West Coast modern dance company at the time, was selected as the resident company. For this group, private fund-raising was a measure of civility, providing access to a place in the center. The company was given seed money from the CRA/LA in the form of a $5.5 million loan for facilities but was required to raise matching funds to demonstrate it was financially stable. When it failed to raise the funds, the CRA/LA support was revoked and the dance company never moved into California Plaza. However, the second office tower where the dance gallery would have been housed was already built by the time it became clear that it was unable to raise the necessary funds. Artist dressing rooms intended for the dancers are used as offices for Grand Performances' technical staff and summer interns. "Artists' Entrance" signs still hang in the service hallways that are now used for mail delivery, garbage disposal, and as a shortcut from the production space to the Plaza.

A space for public performances marked a culmination of the Bunker Hill Urban Renewal Project's call for a place for "cultural gatherings of many kinds" and "a plan of land use of great benefit to the people of the entire Los Angeles metropolitan area" (CRA/LA 1959:23). Part of a wider Central Business District redevelopment strategy, the California Plaza development was one among several places where the CRA/LA could use public performances to bring people into downtown public spaces (Garden Theatre Festival, Box 62). As mandated by the CRA/LA, public programming in the Central Performance Plaza would be provided by the developer, with programming and budgeting supervised and approved by the CRA/LA. Representing and constituting "the people of . . . Los Angeles," the development of the content of these performances and the design of the public space went hand in hand and were the central means by which an urban public was crafted for and by California Plaza.

The first plans, presented by the winning development team, described a space for nearly every kind of event, including festivals, concerts, filming television programs, and visual art exhibitions. The imagined public of the Plaza was vague and broad, categorized by diversity of work, transportation, and taste. The developers imagined the public space as a vibrant, festive place, bustling day and night with social activity and performances.

> The plaza of Angel's Park is a public forum, offering a natural stage for civic festivals, concerts, dance or televised shows. It offers a perfect opportunity to develop the wealth of budding talent in Los Angeles. During the day it is an ideal setting for informal concerts and outdoor exhibitions supported by cafes and shops. At night it changes character. Innovative restaurants and cabarets provide excellent alternatives for drinks or dinner to those attending performances under the canopy on Angel's Park plaza or to patrons from the nearby Music Center. (Bunker Hill Associates 1980:32)

The CRA/LA returned the first proposal to the developers for refinements and a second plan was drafted. "Plaza for the Performing Arts" added "the populace" to the mix of consumer segments, which consisted of Los Angeles' minority groups: Hispanic Americans, Asian Americans, African Americans, children, and women (Bunker Hill Associates 1982:7–8).

Subsequent to this initial articulation of social diversity and with his control over the design process slipping, architect Arthur Erickson drafted a reactionary plan that sharply redefined the public of California Plaza as an elite arts audience. In 1985 he presented a plan to the CRA/LA that stated, "One of the main objectives of this space is to provide entertainment of various types and times to the public of California Plaza." This public was defined through programming. Informal performers such as jugglers and street musicians, noontime concerts featuring string quartets and boxers preparing for competition, and evening performances of the Los Angeles Philharmonic and rock groups would attract "well-educated, affluent groups which will have an appreciation of cultural events and will have a fair amount of disposable income which can be used to support these cultural interests" (Arthur Erickson Architects 1985:15–16, 14).

Two years later, as building commenced, a plan titled "The Arts in an Urban Garden" returned to an emphasis on social diversity, describing a varied space with different kinds of performances where residents of and visitors to Los Angeles would help make it known as not just office towers but "an urban attraction to visit and enjoy during the day, on nights and on weekends" (Bunker Hill Associates 1987:5). A plan for Angel's Flight Productions, submitted by Aaron Paley in 1989, outlined explicitly multicultural programming, relating it to a city now described as diverse. This plan provided the basis for what would become Grand Performances.

Models for a public concert series in California Plaza were provided by citywide festivals and other downtown concerts. The former made multicultural programming a viable and in some ways a necessary choice, while the latter used the arts to attract Angelenos to spaces either unoccupied or occupied by undesirable populations. These programs, and their surrounding agendas, sponsorship, and controversies, laid the foundation for Grand Performances. The Garden Theatre Festival was initiated in 1970 as a small theater festival in a backyard near the University of Southern California; within the decade it had become a multidisciplinary, multicultural event in the city's Barnsdall Park with programming intended to reflect a diverse Los Angeles. In 1979 its director was hired to present concerts in downtown Los Angeles with the explicit mandate of "activating" and "cleansing" urban space.

Earlier that year public performances in Pershing Square, presented in conjunction with the Garden Theatre Festival, had attracted an estimated six thousand people to downtown Los Angeles. As Whitt warns, "the arts attract people to downtown, but not just any kind of people" (1987:28). The director of the Garden Theatre Festival explained, "Pershing Square was undesirable. It attracted a lot of homeless because it's close to Skid Row undesirables. They'd closed the restrooms because that was where the homeless urinated. And they brought me in because they wanted to change that." Adjacent businesses that did not already support the arts contributed to the performances after they were shown how the arts could, the presenter said, "clean up the front lawn."

A proposal for the Festival in the Square explained that Pershing Square "has a strong reputation as the 'park that people run through' as the 'residents,' as it were, are not a popular element with the working community" (Garden Theatre Festival). As the authors of the proposal maintained, "We felt this was an opportunity to put the arts to work and test the theory of revitalization." According to subsequent discussions, the project seemed to work.

On the whole . . . we had very little trouble, and had fabulous coop-
eration from the police . . . but that "element" of people did play a
significant role in the response to the night programs. The first eight
weeks were so successful that the "residents" almost faded into the
background, but as people began to get comfortable with the programs and the numbers started to dwindle, the "residents" began to
be very apparent and started an uncomfortable cycle of reinforcing
the negative aspect of the park as safety in numbers began to
decrease.

In "activating" public space by attracting people there, downtown performances mobilized one group to "cleanse" urban space of another in order to
ultimately support business interests.

During the 1984 Olympics, forty-five days of consecutive concerts were
held in Pershing Square to, as one presenter said, "make it look like there
was something going on." Elsewhere, downtown free concerts supported
by corporations were used to draw audiences to otherwise underutilized
privatized public spaces. Weekly noontime concerts on the Bonaventure
pool deck sponsored by Arco[5] drew five to six thousand people to performances by popular artists such as Dianne Reeves, the Yellow Jackets, Queen
Ida, and Bo Diddley. Arco stopped sponsoring these concerts around the
same time free concerts began in California Plaza.

The first free lunch-hour concerts were held in California Plaza's Spiral
Court in 1986 and were produced by Community Arts Resources, Inc.
(CARS). In 1989 its director, Aaron Paley, was hired to draft what became
the ultimate plan for programming at California Plaza. A veteran of the
1984 Olympic Arts Festival, Paley had most recently worked on the first
Los Angeles Festival held in 1987, a triennial event he continued working
with until 1993 as organizer of the Fringe Festival. The Los Angeles Festivals
took up the mantle of the Garden Theatre Festival in their orientation
toward the city as a whole. While the 1984 Olympic Arts Festival emphasized bringing important international programming to Los Angeles and
putting the city on a global cultural map, the Los Angeles Festivals (1987,
1990, 1993) turned increasingly to the city itself as the source of performers.
Organizing the 1993 Festival around the diversity of the city, its director
urged "people to really live in their city. Encounter it instead of avoiding
it" (Peter Sellars, qtd. in Kreiswirth 1993:K16). CARS's project is founded
on the multicultural, citywide outlook of the latter Los Angeles Festivals

(Cheng 2002:71–76; Davis 1992:80–81; Kirshenblatt-Gimblett 1995; Lowe 1996). Today it produces, among other things, an annual festival in Santa Monica; Yiddishkeit, an annual Yiddish festival; and public events at the Getty Museum.

Of all the proposals put forward, Paley's was the one most directly implemented for the concert series in California Plaza, shifting the emphasis of the planned performances firmly toward multicultural arts. A five-year plan for Angel's Flight Productions emphasized "affordability, accessibility, usefulness and vitality" in order to serve the needs of local arts organizations, especially multicultural arts groups (Paley and Slavin 1989:2, 3–4). Drawn from a city defined as diverse, the public of the Plaza was imagined to represent the diversity of Los Angeles. Programming was oriented toward the people of a diverse and segregated Los Angeles, "from tourists to office workers, from Watts to East Los Angeles, from Latin Broadway to Chinese North Broadway, from the Valley to West Los Angeles" (Paley and Slavin 1989:7–8). By attracting residents of segregated and dispersed neighborhoods to a shared public space to participate in a common experience of public performances, programming for this public supported the project of making California Plaza in downtown L.A. the "center" of the city.

Following the submission of this final plan, Michael Alexander was hired as programming director of public events in California Plaza. Alexander had joined the Cultural Affairs Department (CAD) of the City of Los Angeles as director in 1988. As director of CAD, Alexander presented free concerts in the Civic Center Mall around the Triforium, by then in disrepair. Dedicated in 1975, the Triforium is a kinetic sculpture with electronic sound and light features. Afraid that audience members might fall into the pit left by a reflecting pool that had been drained after it had leaked into the parking lot below, Alexander also wanted to find a performance space that could accommodate more people. "My concern," he said, was "that the city of Los Angeles, as the number two city in the country, needed to have a more substantial summer program than one that would only accommodate two hundred people. It struck me as just an inappropriate use of public funds to be producing something that was playing to such a rarified group of people in this city of three million people." In 1989 the city produced evening concerts in the Spiral Court while CARS produced lunch-hour concerts. Later that year Alexander applied for the job of programming director of the fledgling performance series.[6]

Having grown up in Los Angeles, Alexander was a veteran of community theater, had performed as a clown, and had managed the Aman Folk Ensemble. He credits his investment in multicultural and international programming and civic performance to this experience. Heavily influenced by the ideals and approach of the Garden Theatre Festival and Los Angeles Festivals yet aware of the pitfalls of risk in the context of its corporate setting, under Alexander Grand Performances continues to walk an artistic line between a populism that reaches out to a city marked by diversity and a corporatism that accedes to the demands of its venue and major supporters. Its project entails bringing the "right" people to California Plaza, who will help constitute a multicultural audience but not make class apparent, who will "reflect the diversity of L.A." but not express divisive political views.

Organizing its presenting around an imagined general public of the city, interests of the corporate plaza, and grant guidelines of government arts agencies, Grand Performances enacts a state project of multiculturalism. As such it sustains a foundational contradiction of multiculturalism as both political and depoliticized, as a site for the production of forms of belonging wrought from exclusions. Multiculturalism reflects conditions of neoliberal capital that have contributed to the transformation of the sociogeography of cities and turned identity into a consumer product at the same time as it presents the promise of new forms of rights, access, and belonging (Dávila 2001; Kymlicka 1995; Lubiano 1996; Povinelli 2003; Taylor 1994). Complicit in the production of ethnicized subjects by capital and the state and the performance of delineated identities by those subjects, multiculturalism also creates the conditions of possibility for those identity categories to provide the terms for politicized resistance. Ethnicity as a marketing category and a means of making minority group claims on the state is imbricated with belonging, which might be organized around capitalist consumption as well as identity politics that are outside the logic of capitalism.

Grand Performances navigates between multicultural arts projects as political, ethical engagements on the one hand and the folding of multiculturalism into the interests and constraints of capital on the other (Lowe 1996; Yang 2008). The nature of nonprofits sustains this stance. Nonprofits exemplify the privatization of public services in the United States, as reflected in their growing numbers and increasing role in administering social welfare. Yet nonprofits stand between public and private, existing as both at once, mediating and blurring the line between "public" and "private" rather than simply reflecting the privatization of the public. Intended to provide public goods and services, nonprofits must compete in the marketplace for legitimation and funding. At the same time, referred to as the "third sector," nonprofits are tax categories that emerged out of early American voluntary organizations (Hall 1992; Hammack 1998). They are closely linked to government through their tax-exempt status, by offering tax relief for donors, and as the only kind of organization eligible for government arts grants.

Grand Performances' nonprofit status shapes its publicness in terms of what makes it "public" and what it offers as a "public good." For Grand Performances' director, the transparency of nonprofits' finances makes it more public than government agencies. Defining free concerts as a public good insofar as they are not profit driven, Plaza Commons, Inc. (PCI; the original presenting organization), in its first attempt to gain nonprofit status, was denied because it was deemed to be a for-profit business, earning revenue primarily from the parking garage.[7] In 1996 Friends of California Plaza Presents was formed as a "nonprofit public benefit corporation" and granted nonprofit status by the state of California.[8] In its successful application, Grand Performances (then Friends of California Plaza Presents) stated that its public good was providing performing arts as education. As its application for nonprofit status argued, "The specific purpose for which this corporation is organized is to educate the public in the performing arts through the presentation of public performances at California Plaza in downtown Los Angeles."

Art as deployed by civic performance figures multiple meanings of public and private. The public-private divide is largely framed as a presumed opposition between art and capital. This opposition figures Grand Performances' public concerts as existing at the crux of a market-driven city and a logic of art as outside of and antithetical to capital. In putting art in the service of a public-private dichotomy, public concerts shape the meaning of art through intersections of programming, funding, and marketing. Art is not inherently public. Rather, art provides a package for a range of public goods that include the idea of art as a public good and extra-artistic ends such as community, multiculturalism, and cross-cultural understanding. At the same time, art is often privately funded and part of multiple private practices and institutions.

Americans have long debated whether the arts are a public good worthy of state support (Banfield 1984; Benedict 1991; Starr 1987). These debates have asked what kind of good the arts are, what the role of the government is in funding arts, and what kinds of arts the government should fund. Even as municipal support became a defining feature of public concerts, early examples show there has always been a range of funding. And when municipal governments funded free concerts, this support did not come automatically or unquestioningly.

In its use of art, Grand Performances operates within a set of private-oriented effects of neoliberalism at the same time as it proffers public goods that are opposed to profit: free art for everyone, payment for artists, and community building. The logic of the arts creates a space for the imagined reflection of urban diversity. Its social consequences consist in drawing "minority" audiences originally from the country of origin of the performers, who feel included in the civic body by seeing performing arts from their country, who help signify the diversity of the urban public by their presence at the performance, and who are members of the moral community of the public concert audience. The market logic, on the other hand, affects presenters' choices about which countries' artists should be presented and which artists will be programmed, as well as marketing and development activities that are in turn related to other local cultural programs.

As a nonprofit, Grand Performances became eligible for government funding at national, state, county, and city levels as well as for foundation grants; it was also able to offer a charitable tax write-off for corporate and other private donations. Today roughly half of Grand Performances' funding comes from corporate sponsorship, government and philanthropic grants, and audience donations. At every evening performance during intermission, staff, board members, and other volunteers walk through the audience carrying five-gallon red buckets, decorated with colored lights to make them more visible in the dark. A volunteer commented on how it was similar to "tithing" at church. Like tithing, red buckets help create the audience as a moral community, active participants in making the free concerts possible. In pre-concert and intermission comments the director expresses a dependence on audience members' contributions in order to "help us get through our season," urging them to follow the invented example of last year's audience who "averaged three dollars per person." In actuality, audience donations average around fifty cents per person and usually add up to around $25,000 over the season. One summer a man asked audience members for money for himself when volunteers brought red buckets around at intermission. Grand Performances' staff asked him to stop, and by the end of the season he would tell them, "I'm not asking for money now."

Nonprofit status supports Grand Performances' multicultural project by bringing together the aspirations expressed in the final programming plan and available arts funding. Through nonprofits, the symbolic capital of multiculturalism can be transformed into economic capital. With world music and international programming already a foundation of the performances under Michael Alexander, the board first discussed diversity in the context of available funding and desirable board members. After becoming a nonprofit, Grand Performances shifted its emphasis from noontime to evening performances and presented a more explicit multicultural project. The board stated, "We want to have a substantial arts program which is multicultural and varied (music, dance, etc.)." An arts consultant advised the organization that arts funding sources are "looking at the impact an arts program has on the community. Funding is generally more available for programs with significant community impact rather than for 'Cultural Palaces' (such as the Music Center)" (Board Minutes, December 5, 1996, Grand Performances' files).[9]

Grand Performances structured and solidified its multicultural work by forming partnerships with other local arts organizations oriented toward specific racial or ethnic groups such as the Watts Towers (African American), Self Help Graphics (Latino/a), and Plaza de la Raza (Latino/a). The first proposed mission statement, drafted at the beginning of 1997, used language that began to reflect what the staff now considers the organization's multicultural project: "The mission of California Plaza Presents is to bring together, in a unique urban venue, the diverse people of Los Angeles through performing arts that reflect those communities." In this way they produced a meaning of multiculturalism that looked to the city for its content. Desired board members—who would help create a multicultural audience at Grand Performances' concerts—were "top layers of all ethnic communities; liberal; people interested in a diverse community; the educational community" (Board Minutes, January 14, 1997). By 2003, with the early goals seemingly met and institutional stability a greater concern, emphasis had shifted to finding board members with money or connections to money who could provide financial support for the free concerts.

A name change shortly after the nonprofit was formed discursively connected the free concert series to wider trends of downtown development.

The name change was intended to both broaden the geographic scope of the concert series and help the organization obtain funds from sources beyond California Plaza: "The title and approach of Friends' publicity could be broadened from just California Plaza to 'Grand Avenue' in order to attract greater support from neighbors of California Plaza" (Board Minutes, October 21, 1997). The new name, Grand Performances, met the ideal of being "memorable, exciting," and able to "spark one's imagination" (Board Minutes, November 6, 1997). The name also had "a geographical component," associating the concerts with Grand Avenue, which, as the home of the Music Center and the Museum of Contemporary Art, was already a destination for cultural consumption. The name was officially changed in 1998, with a byline added to the stationery that, evoking the role of the arts in attracting the "right" people to an area, stated, "because people are returning downtown."

Public Bodies

Although the final Plaza hardly resembles what one imagines from the ebullient descriptions of the plans, elements of their designs can be found in the space that was built. Today, the central performance area doubles as a water feature, where shooting jets cascade into a waterfall that flows down a graduated terrace into a reflecting pool. When turned off, the surface between the rear shooters and front pool becomes a stage. The audience sits across the water from the performers. On the lower level plastic chairs are tied together in rows, facing the performers in three sections around the perimeter of the water. The middle section, in front of the sound booth, is reserved for donors, board members, press, and friends of the artists. Upstairs, audience members sit on granite benches built into the Plaza. People stand to watch over the heads of those seated, and behind them others dance. For smaller events, a side area called the Marina Pavilion is used as the stage, and the audience sits on benches created by the granite amphitheater. Though spontaneous events were once imagined as a possibility, today performances at California Plaza are always planned, with only one taking place at any given time.

The planning process wavered between desire for openness and fear of the public, ultimately leaning toward the latter with ensuing results in the built environment, security, and event planning. California Plaza, in particular during public concerts, maintains enough elements of civic ideals to

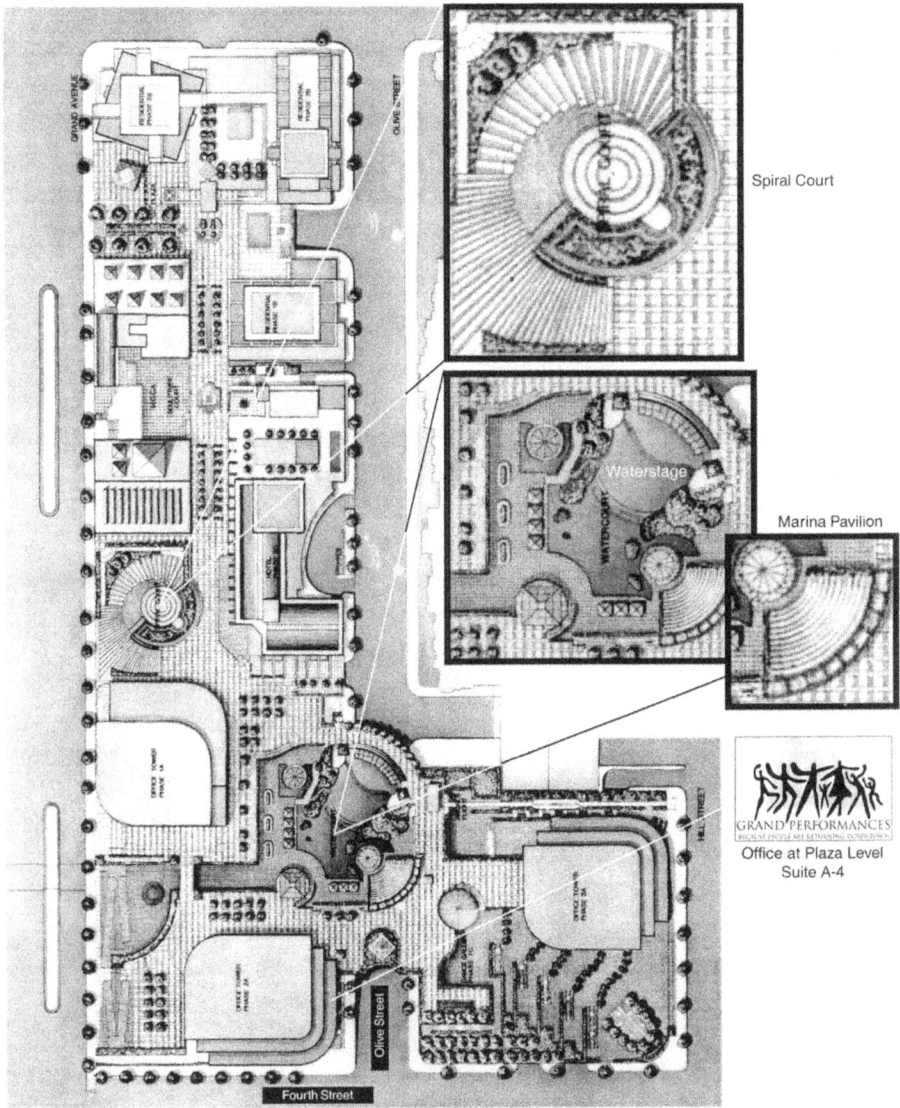

Figure 3. California Plaza's performance spaces. Grand Performances has overlaid its offices and performing spaces on a projected plan that included three office towers, the last of which was never built. Courtesy of Grand Performances.

be recognized as public space and for users to make claims on that basis. Even a critique of California Plaza as failed public space depends first on a recognition of it as public space. With a history of negotiations over the nature of the public space sedimented in its built environment, the publicness of the space and people is continually created through use. Practices of visitors, security measures, and use regulations help shape the public of California Plaza by disciplining bodies of included members of the multicultural public and actively excluding others. Disciplinary measures are both organized around and put pressure on foundational tenets of liberal democracy such as openness, tolerance, difference, and dissent. Marking while revealing the limits of the publicness of the Plaza and performances, these tenets are also invoked to make claims on its publicness.

The only sign alerting a visitor she is entering California Plaza is a small plaque on the ground marking the divide between the public property of the sidewalk and the private property of the Plaza. Evoking openness, the lack of posted verbal rules in California Plaza suggests a general "fit" between desired user and social "order" in the Plaza, with a user whose practices and accoutrements are considered proper for the place. Her presence in the space is naturalized, as her habitus, or "bodily hexis," is continuous with the status quo of the "established order" (Bourdieu 1984:427).

The openness to desired users belies the ways in which undesired users are excluded, putting pressure on ideals of tolerance for difference. Class is excluded as a category of difference. Grand Performances' staff decide not to invite homeless children to concerts because it will create a "class-noticed situation;" they talk about how the janitors sometimes come back to California Plaza at night for the performances with their children "all dressed up" and thus able to "blend in." Those who are visibly homeless or panhandling are asked to leave for fear that they will drive part of the diverse, middle-class audience away or force them to face existing social problems. The exclusion of class is necessary for the creation of a multicultural audience in which identity is serial, measured by differences in form and content of color and "culture" rather than economic conditions that reflect social inequalities. The requirement for membership in the multicultural public is the display or performance of ethnicity or race. Grand Performances' success resides in the recognition of ethnicity or race as constitutive of a multicultural whole by at least some of those who are present.

Security guards patrol the common area of California Plaza on foot, looking for forbidden activity. Wearing suits and ties, they blend in with the daytime corporate users, helping maintain the consensus of privatized public space while they enforce movement through the Plaza. The guards' gaze is extended by omnidirectional security cameras hidden under half-domes of dark Plexiglas located throughout the common area. Two guards sit inside a small room on the ground floor of one of the office towers watching the monitors that show the view of each security camera. If the guards see something deemed suspicious they will pause the automatic rotation of the cameras, holding an image on a monitor that shows the view from the camera capturing the activity. One guard chuckled at the memory of moving the camera away from a couple having sex at the end of a dark corridor, unaware they were being watched. If the suspicious activity is a person walking through the Plaza, the guard uses her fingers to push the arrowed buttons that control the movement of a camera, turning it to "follow" the person and picking up his movement with the next camera.

During an afternoon that I spent in the monitor room, California Plaza's only female security guard trained me to "follow" a person on the Plaza with the cameras, as she thought that was the most significant thing I could learn. I clumsily translated the view of the camera onto the arrows beneath my fingers, first turning it the wrong way, then too far off its course before getting a very rough sense of the connection between my fingers and the visual space of the Plaza. Though I practiced on a businessman dressed in a suit, the security guards generally use the cameras to follow someone who has visible traits that mark him as homeless: unkempt, dirty hair, a blanket wrapped around his body, perhaps missing a shoe or two. After following the person's movement, the guard watching will call a guard patrolling the Plaza on foot, telling him there is a "guest." The guard will ask a person panhandling aggressively to leave the Plaza. If the person is unkempt but not disturbing anyone, the guard will stand three feet away, following until the person feels uncomfortable enough to leave the Plaza. At performances, homeless people are generally asked to leave only if they are panhandling or creating a disturbance.

The disciplining of bodies in the Plaza extends to concerts. Such disciplining is central to the production of consensus in privately owned public space. A proper performance by audience members is necessary for the concert form to work and for those present to be constituted as civic subjects and members of the public. While audience members will in general "act" their part correctly by listening to the music, watching the performer, telling a noisy neighbor to be quiet, and applauding at the end of pieces, their movement is micromanaged to limit the claims people can make on the space and to create order at the event. Discipline is maintained through rules, security measures, and the invocation of social norms by fellow audience members. The interconnection of these areas of legitimation yields a general consensus such that "rules" are less visible, and public space is constituted as a space of consensus rather than dissent, a space of civility that enforces particular modes of bodily comportment. Although there are legal frameworks behind these regulatory measures, for the most part they are unknown and rarely raised, and the rules themselves, rather than questioned, are enforced by all present.

During concerts, a sign is posted listing Grand Performances' event policies. These policies, rather than specifically disallowing practices, maintain consensus by asking patrons to comply, for their sake and that of others.

> Grand Performances reserves the right to hold seats at its discretion.
> House opens 2 hours prior to performance.
>> Seats may not be reserved prior to house opening.
>> One person may reserve a maximum of 6 seats.
>> One person must be present to hold their seat reservation.
>> Blocks of seats indicated by blankets, tape, etc., will not be honored.
>> Please keep aisles clear at all times.
>> Please use plastic, paper or aluminum containers for food and beverages.
>> Please move away from the seating areas if you wish to smoke.
>> Please use ash cans to dispose of cigarette butts.
>> Please move away from the seating areas to make phone calls.
>> Seeing-eye or other assistance animals are welcome at California Plaza.
>> Patrons should not bring their pets to performances.
>> *A full list of Grand Performances policies is available at the Information Booth.*
>>   *Thank you!*

These semi-official security guidelines and event policies are the only rules made visible to the public. The "full list" of policies expands on these rules, asking audience members not to drink red wine as it might stain the granite and requesting that people with reserved seats arrive by 7:45. Additional rules include not passing out flyers on the Plaza (people are allowed to put their material at the information booth) and not filming without permission from Grand Performances and the artist. Problems and "concerns regarding the safety of the audience and the enjoyment of the performance should be reported to the marketing cart to be passed on to GP [Grand Performances] management or handled by security as required." The differences between the two sets of rules depend on whether "problems" are considered widespread or limited; general problems require a notice that is visible to all audience members, while limited ones can be regulated by staff members or security. The existence of the full set of written rules is used to authorize these enforcements if contested by the perpetrator.

Performance regulations are supported by the 1981 California Plaza Disposition and Development Agreement (DDA), which asserts that Plaza use guidelines are at the discretion of the developer (DDA, Exhibit C, CRA/LA 1981:5). While rarely read or referenced as such, they are cited by Grand Performances' staff members when disciplining an unruly audience member or describing the nature of the public space. Along with controlling users' activities, these rules made it such that only space for administered events was written into the nature of public space. In the name of the sensibilities of an abstract "Permittee," users are not allowed to do anything that might constitute a performance, such as "Parade, rally, patrol, picket, demonstrate or engage in any conduct that might tend to interfere with or impede the use of any of the Common Area by any Permittees, create a disturbance, attract attention or harass, annoy, disparage or be detrimental to the interest of any of the establishments within the Center" or "Use any sound making device of any kind or create or produce in any manner noise or sound that is annoying, unpleasant, or distasteful to any Permittee."

Use of the space is also controlled by less explicit devices that help audience members "perform" consensus. Maintaining order, they discipline bodies into concert audience members qua civic subjects. At evening concerts, chairs are laid out in rows and tied together to keep them from moving. Stanchions are placed around reserved seating sections and a staff member monitors access. Crew members put yellow tape on the ground to create fire exits and walkways. The effectiveness of designations regarding where one is allowed to stand, sit, or move through must be enforced by people. A security guard stands at the edge of the yellow tape, admonishing audience members who pause to catch a glimpse of the performer to move on.

Though guards and regulations put pressure on the publicness of the space and performances, they also maintain a space of consensus in which they are generally not questioned as such by users. However, the relative publicness of the performances is raised at its limits. The fact that the concerts are free posits an ideal of equality that is undermined by reserved seating, which stratifies based on financial contribution through the reward of a superior consumer experience. The stratification created by reserved seating is spatialized in California Plaza. For every evening performance, seats in the front, middle section of the Lower Plaza are reserved. At performances to which many special guests are invited, seats are also reserved on the benches on the right side of the Upper Plaza. The seats in the Upper Plaza are considered the best for watching dance, as the four-dimensionality of the choreography can be seen most clearly from above. Before a performance, the reserved areas are cordoned off and a staff member, holding a clipboard with a list of names of those with reserved seats, a pen for crossing off the names of those who have already arrived, and a radio in case help is needed, is stationed at each entrance.

Grand Performances' returning audiences learn quickly that in order to find seats they must arrive early. People begin to arrive an hour or two before the performance is scheduled to start, and for a popular performer the seating areas might be filled within half an hour of the first audience members' arrival. Those who come late usually have to stand. However, audience members with reserved seats do not arrive early, as their seats are saved for them. This can create a situation in which all the seats are filled except the best ones. People arriving at this point are often dismayed, asking whom the seats are for and why they are reserved.

At a Friday evening performance I worked at Upper Reserved. Early in the evening a man became quite angry at the fact that the area in the Upper Plaza was reserved. "This is a public place! They can't reserve these seats. I'm sitting here," he announced loudly, eliding a meaning of public as general, nonhierarchized access and ownership by "the people" into the right to claim private ownership. While "they" do not have the right to reserve seats, he, self-defined as a member of the public, could register a claim to space with his body rather than leaving the spot for another. His girlfriend was embarrassed and tried to get him to move. "Be nice," she wheedled, begging him to consent to the arrangement. Having had no suc-

Figure 4. Tape marking walkways for a performance, and a guard looking out over downtown.

cess in convincing him to move, I asked a member of the crew to help; he spoke calmly to the man, explaining the cost of producing the performances and telling him that seats are reserved for donors. "But I donate in the red bucket," the man grumbled, before he moved to the other side of the stanchion. Others, seeing the empty seats, walked past, not even noticing me standing there, and sat down. I asked them to move, explaining that the seats are for donors who, as I had been told to say, "help make these free performances possible."

Grand Performances' staff members argue that part of their aim is to make everyone feel welcome at these downtown concerts and that they want people, in particular those marked as "minority," to have a sense of ownership of the event and of their city. Many public concert audience members exhibit a sense of ownership after having been to the place over and over and claiming particular areas as "theirs" in which to sit. By claiming seats they constitute themselves as members of the public for whom the space and the event are intended at the same time as they "privatize" those particular spots, claiming individual ownership of them for the duration of the event. This ownership could only rightly be usurped by another audience member who arrived earlier and happened to sit there; for the institution to mark the seats as privileged appears to audience members to privatize the space too blatantly and obviously, erasing, or at least calling into question, an ideal of equality implicit in this public space.

Conclusion

The multicultural urban public now recognized in the public concert audience is predicated on a history of city- and civic subject-making informed by interests of capital and modernist notions of cleanliness and proper living. Sounding city space, civic performance's modes of inclusion are wrought out of prior exclusions that structured both bodies and geographies. Urban renewal helped create a segregated city that is perceived as needing to be brought back together in ephemeral sites of pleasure such as free concerts. Those displaced were promised new, modern housing in another part of the city, in order for the general population of the people of Los Angeles to benefit from increased property tax revenue and a public space that could be used by everyone. The arts are used to draw the segregated city wrought out of urban renewal back together in the center, uniting an urban public that reflects the diversity of Los Angeles. In attending performances on the land of urban renewal, residents of the neighborhoods to which Bunker Hill's populace was dispersed are brought back together under the auspices of a multiculturalism that only refers to that history as evidence of L.A.'s diversity, flattening a time span of neighborhood change in space.

Membership in the urban public is made through access to, ownership of, and presence in the center. Organized around diversity, togetherness, and consensus, downtown civic performance enables a multicultural public to emerge as a normative urban public, supporting senses of belonging and recognition of a diverse city. At the same time, the bases for Grand Performances' diverse audience to come together downtown from around the city reflect and re-create multiculturalism's tension between the particular and the general. This tension is expressed spatially and sonically. Spatially, it is manifest in the relation between segregated neighborhoods and a "neutral" downtown. Insofar as members of a multicultural urban public come from racialized neighborhoods around the city, making downtown neutral always depends on a relationship to a surrounding city understood as segregated. By moving through urban spaces in which they are alternately marked and abstract, urban subjects create a relationship between the normative and the marginalized that is sedimented in a real and imagined racialized geography. As sonic expressions of multiculturalism, Grand Performances' events draw audiences with affective, identity-based connections to particular musical forms to concerts where they are part of a collective recognized as diverse. Grand Performances' concerts aspire to constitute a unified public of the audience, allowing for the expression of differing views while encouraging consent for a notion of multiculturalism and tolerance of difference.

2

Mapping a Metropolis in Motion

Celebrate the world . . . Celebrate L.A.!
    —Grand Performances brochures, 2001 and 2002

Because it's downtown, it attracts everyone from all
places—Westside, East Side, North, everywhere. Because
there are so many different areas, the Pico area, Echo
Park, South Central, as well as West L.A. can all go there
and be in the same place.
    —Grand Performances audience member

When I entered Grand Performances' office to begin fieldwork, the director handed me a well-worn copy of *The Ethnic Quilt: Population Diversity in Southern California* (Allen and Turner 1997). They were interested in the information in the book, he explained, for fund-raising and marketing. I leafed through its pages, stopping to look at maps showing the spatialization of 1990 census data. A blue map with the caption "Persons in Poverty" reflects a concentration of purple in and around downtown (31). A tan map reveals that people of Iranian Ancestry—as denoted by the book—live mostly around Beverly Hills and Brentwood in West L.A. (59).[1] Areas in which Blacks compose more than three-quarters of the population make a jagged line from Crenshaw south to Inglewood east to Watts and south to Compton (62). Hispanics are broken down into their countries of origin; maps on opposing pages distinguish that Mexicans live east and south of downtown while Guatemalans are concentrated northwest of downtown (100–101). Chinese, who compose between half and three-quarters of the population of Chinatown, otherwise largely live in areas to the east such as Monterey Park and Hacienda Heights, though some Chinese live in nearly every part of the county other than counties that are predominantly Black.

One map shows movement. Titled "Major Shifts in Ethnic Populations After 1940," it uses arrows to represent the direction in which various ethnic populations moved (Allen and Turner 1997:51). Everyone moved away from downtown, which, circled by tails of arrows of various colors, is a center of focus on the map. Large white arrows outlined in black show a centrifugal spread of the White population from a circumference around downtown. Green arrows indicate the Mexican population's stability and movement from downtown and East L.A. eastward, while pink arrows trace the trajectory of Black migration from Central Avenue to adjacent areas south and west, and farther away toward the town of Pacoima in the San Fernando Valley. Chinese, in orange, remain in Chinatown while having also moved east and northeast. Though the rest of the maps represent a particular, fixed moment of residential patterns, the text emphasizes change, as each chapter, focusing on a broad ethnic category, provides a history of its subgroups' arrival in Southern California. In constructing a vision of the city in which discrete demographic categories are linked with territory (Wood 1992), the book is useful for Grand Performances in its effort to identify areas of population density of ethnic groups. Yet the representation of residence fixes a city in motion, a city in which people and sounds circulate within and between neighborhoods. This is motion that Grand Performances also depends on and invokes in presenting performances downtown and in advertising them to people around the area so that they will move through the city to attend a performance, where they might dance or otherwise be moved by the music.

Grand Performances sounds the city in and through its rhythms (Lefebvre 2004). As a media producer, it taps into sonic and social circuits of a metropolis in motion (Ginsburg, Abu-Lughod, and Larkin 2002:17). Radio advertisements, word-of-mouth recommendations, television announcements, and brochures with visual and textual representations of musical performances expand the sonic range of public concerts. A process that is dynamic and mobile, the circulation of information about performances articulates the urban rhythms of media in everyday life, their use marking everyday temporalities and geographies. Like the "imaginative texts" of urban cinema, civic performances become "spaces of association" that "take on phenomenal lives" (Larkin 2008:250) through their relationship to the city, to people's imaginations of that city, and to their connection to media that map onto and imbue urban imaginings and everyday life.

Brochures travel across the city to arrive in mailboxes, perhaps of people who have since moved to other neighborhoods. Radio waves carry sound that becomes music in a car's stereo as it moves with the flow of traffic. A radio station's broadcast range is instantiated in practice by people listening or watching. These soundings intersect with movement to the concert from neighborhoods, rhythms of percussion, sound, and people, the circulation of global capital that built the plaza and supports the performance, and histories of rhythms of others before in their movement to Los Angeles and around the city. Media that map neighborhoods might be in Spanish or Mandarin as well as English, might be directed toward diasporic Iranians or African Americans. Shaping the global city, transnational media from outside the United States or made locally for a diasporic group mark a coalescence of cultures, media, and capital in Los Angeles (Appadurai 1996; Ginsburg, Abu-Lughod, and Larkin 2002:14).

Categories of centrality and sprawl organize Angelenos' navigation through their city even as they serve as general analytics of urban scholars (Keith and Pile 1993; Tajbakhsh 2001). In its use by Grand Performances, *The Ethnic Quilt* becomes part of a wider sensibility in which Los Angeles's diversity is inextricably linked to geography in fact and imagination. Constitutive categories of multicultural diversity are attached to urban neighborhoods that serve as code words for identity. Grand Performances' audience, in purportedly reflecting the diversity of Los Angeles, is understood by those who are part of it and those who help shape it as a spatialization of identity that brings the city's neighborhoods to the downtown corporate plaza. "L.A. is without a majority population," Grand Performances' director told me later. "Everyone comes to Grand Performances, people of all ages, races, income, and geography." To create this audience, Grand Performances' use of marketing presumes and produces anew an inextricable linking of identity and neighborhood.

Grand Performances relies on a certain perdurance of identity categories and their relationships to cultural expression and place that have been solidified through a long social and philosophical history. In constituting multiculturalism by bringing discrete national, racial, and ethnic groups together for a performance, such categories are fixed momentarily by their correlation with media and geography. The bases of this momentary fixing reveal at once the instability of identity categories and their spatialization:

the openness of media allows for consumption by a range of demographics, media use is connected with other practices that may or may not be organized around identity, and racialized geographies shift over time. And though census figures might provide legitimation for Grand Performances' efforts, their terms are continually created and undone in practice, as an imagined fixity is put into motion.

Inhabiting the city, one moves through and performs the city (de Certeau 2002; Roach 1997). For the most part Grand Performances' audience arrives by car, though some who live in lofts in the Historic Core might walk up the hill. A resident of an artist colony in Lincoln Heights carpools with friends or takes the bus. In 2002 I was living in a loft in the Artists District just east of Little Tokyo on the eastern edge of downtown. I could reach California Plaza by taking the downtown DASH bus, riding my bicycle, or walking in just over twenty minutes. Walking took me through Little Tokyo, a corner of Skid Row, the north end of the new loft developments, across Broadway with its feel of Mexico City, through Grand Central Market where I might buy fresh squeezed juice or mole paste for dinner, and up the steep steps next to Angel's Flight, the renovated funicular that famously brought passengers up and down Bunker Hill every day before its removal during urban renewal but now rests at the bottom of the hill after an accident several years ago.

The previous summer, I was in L.A. for a month and had a sublet in Venice Beach steps from the ocean. Taking the bus downtown every morning and returning to Venice in the afternoon provided me with an experience of Los Angeles that was key to understanding what was at stake for Grand Performances in its orientation from a downtown location to the city as a whole. The first day I arrived in Los Angeles I did not go to Grand Performances' concert because I had been warned not to take the bus at night as it was unsafe to wait downtown. When I took it, I discovered that the bus picked me up a block from my sublet and dropped me off four blocks from California Plaza. It took an hour to reach my destination. The bus started and ended nearly empty. In the middle, it would fill with mostly Latino riders who would get off before we entered downtown proper. The middle-class myth in Los Angeles is that there is no public transportation, or to the extent there is, it will not take you where you want to go. This bus had a direct route from the farthest western point of the city to downtown, supporting a geography of the city made through public transportation routes in which downtown is the hub, even though the density of the social

geography of bus riders is clustered between these two points. Grand Performances' audience does, at times, arrive by bus or subway. California
Plaza's accessibility by public transportation is another way in which it is
deemed open to the public of the city.

Later when I had the use of a car I would tune the radio to KCRW, a
local public radio station, for the long drive across town. Sometimes the DJ
would play a track by a group—maybe from Africa or East L.A.—and
announce at the end that it would be performing at Grand Performances
that evening. Other times, Grand Performances' radio ads would come on
during breaks. One morning the voice of Grand Performances' director of
programming came over the radio discussing daKAH, a hip-hop orchestra
whose recording had just been played. Her conversation with the morning
DJ of *The Global Village* continued, as they played and discussed recordings
of that season's artists: Jimmy Scott, Bembeya Jazz, Ramatou Diakité, Agatsuma, Nati Cano and Los Compañeros, Sinaloese, Kronos Quartet, Hip
Hop Hoodios, Ilgi, and so on. After each recording the director of programming would give background on the artist's style and significance, describing daKAH's jazz inflections, Jimmy Scott's influence on notorious jazz
musicians, the horns in Bembeya Jazz, Nati Cano's work in taking mariachi
from the streets into the concert hall. Public radio stations are equally
invested in a fantasy of connecting people through music. That Grand Performances advertises on public and ethnic radio rather than a popular radio
station—meaning one with the most listeners (Carney 2003)—reflects an
agreement between its aspirations and those of public radio. Moreover, it
posits the public of their audience as one that is not the mass public Adorno
maligned but a self-selecting group with a shared disposition toward music
and the world.

The specific geography of KCRW's demographic was betrayed by the
DJ's urging listeners to "take the trek from the Westside" and a remark
about kids being able to use their "nanny Spanish," the latter immediately
recognized and commented on as a gaffe. Throughout, the DJ and the
director of programming gave general information about the time and location of the performances, emphasizing the fact that the concerts are free.
They discussed parking options, the history of Grand Performances and its
relationship with California Plaza, its status as a presenter, and the range of
programming offered. At those moments, as I sat in my car as it sped east
on the Santa Monica Freeway, I was connected to thousands of others
around the area in cars and houses and offices who, in hearing the same

thing at the same time, became embodied iterations of a broadcast map now manifest in present and future social practices. While I would be working in the information booth at that evening's performance, others hearing the announcement might be remembering the event on their calendar, choosing to attend after enjoying a new song, or ignoring it entirely whether because of a lack of interest in the performer or because they had tuned out the sound coming from the radio.

Geographies of Diversity

The director of marketing sat hunched over an updated version of *The Ethnic Quilt* titled *Changing Faces, Changing Places: Mapping Southern Californians* (Allen and Turner 2002). Scrutinizing the map of the southern part of Los Angeles, he said, "This isn't where South Central is. It should be the area they're calling Watts, south on Central Avenue." A geographic designation had followed the African American population with whom it was associated, as the area previously referred to as South Central had become home to Central American immigrants. Grand Performances' staff proceeded to discuss the ways in which neighborhood names and connotations had changed over the last century: "West L.A. used to be west of Western, now it's west of La Brea. Venice and Santa Monica used to be places you went for the weekend and took the train. Now the beach is part of the city. The wealthy areas were in West Adams." The area designated as East L.A. had moved west, and one staff member's mother had laughed when she heard Los Feliz called East L.A. "because it's not!" Echoing this discussion, at another free concert on the city's Westside, audience members were discussing Watts. One taught elementary school there and commented that all her students were Latino, not African American as commonly assumed. "Really?!?" the other replied with a tone of surprise.

These shared understandings of geographic organization of race and ethnicity carry positive and negative connotations, as neighborhood names become codes for social groups. Around the time of these discussions in Grand Performances' office, the *Los Angeles Times* reported debates over the city's decision to rename "South Central" as "South Los Angeles." While some supported the effort to remove the area's negative connotation, saying it would bring more jobs, others argued that eliminating the term would not "eliminate the stigma" (Gold 2003:A24). The media, reporting on the proposed name change, had, according to residents, contributed to the stigma by using South Central to discuss anything bad that happened "south of the freeway" (qtd. in Gold and Braxton 2003). Though often discussed innocently, correlations between social group and geographic location that circulate in discourse and condition Angelenos' perceptions of their city are echoed and reinforced by publications that map census data (Allen and Turner 1997, 2002), now used for public concert marketing.

In articulating an understanding of its audience and its relation to the city, Grand Performances' staff draws on identity categories fixed by the state through census categories, creating them anew through a complex web of programming, funding, and marketing. As such Grand Performances is an informal space of governmentality, looking to the state for the categories of evaluation that serve as validation of its efforts to reflect the diversity of the city. The census provides a totalizing, "rational" account of demographic categories and figures, which, even if questioned or reconfigured, imbues understandings and perceptions that do not require empirical justification. Recognition of the demographics of the city and audience is built on understandings gained over lifetimes, from personal experience of the city, media, the census, and expert publications, and from a range of other sources of information that help create and define a diverse Los Angeles. Angelenos' membership in a multicultural city is identified according to categories of ethnicity, race, and nation. Grand Performances' own audience demographics are generated from staff observation at events and mailing list sign-ups. Audience members' marked bodies signify existing census categories and staff members' visual estimations provide the figures used on grant applications and reports to the board. In generating categories and percentages, qualitative information is construed in quantitative terms. The production of demographics through visual recognition at performances, along with programming and marketing practices for which they provide an organizing principle and motivation, is an enactment of biopower that helps shape and define bodies as state subjects outside the state proper (Foucault 1978).

Recognition of ethnic groups present in the audience provides a means of evaluation of success or failure in drawing the desired audiences. The Beijing Modern Dance Company's 2003 performances were deemed successful because the audience appeared to Grand Performances' staff to be 50 percent Asian, the highest percentage, they said, that had ever been seen at one of their events. This reflected Grand Performances' success in drawing new audience members and in making connections in the Chinese community in Los Angeles. Grand Performances' staff hoped that these new audiences would return for performances from countries other than their own. Whether or not this happens is in large part unknowable; however, years of talking to audience members provide evidence for Grand Performances' staff that people do return for future performances. The number of mailing list sign-ups at a concert, the only numerical evidence collected, is also considered evidence of having reached their "new" audience segment (for this performance, Chinese Americans), that they are increasing the number of people they are reaching in the city (information primarily used for grant applications that demand evidence of "access"), and that these people might return (because they have indicated that they want to receive the season brochure in the mail the following year).

Visual recognition of audience demographics is supported by mailing list entries from a line requesting the person to "please describe your ethnicity." While fixing ethnicity as a category, at the same time this space reopens demographic categories, guided by and potentially affirming the director's assertion that "race is no longer an absolute." The staff members' discussion of what they would write in this space conveyed their understanding of the relational nature of categories and an aspiration to respond to contemporary circumstances. When asked what he would report, the director replied, "it depends on who's asking. Caucasian, but I might also put Jewish American, or non-practicing Jewish atheist, for cultural events." The director of development said she would simply put "white," and that the hyphen-white labels are a result of whites' feeling left out of new identity labels. This discussion echoed an earlier conversation in which the director of marketing discussed his discomfort with the label "African American." Though from Jamaica, he would not want to identify as such insofar as he would be "carving out a niche" for himself. In the United States he is perceived as African American because of skin color and hair and so would probably report that. When someone raised the possibility of changing the label used in the office to "African descent," he said, "I don't have a problem with that. That's what I am."

While imbricated in wider modes of governmentality that require one to conform to a prescribed category in what Hollinger refers to as the "ethno-racial pentagon" (2000:23), in not providing categories for audience members, the request to "describe your ethnicity" opens a space for identifications within and outside those provided by the state. Moreover, the openness of the request encourages identification rather than refusal, and challenges are directed toward the civility of the question or to the silent state whose categories nevertheless loom large. Selecting—or writing anew—a term that describes one's ethnicity is a process of subject formation that situates a person as a member of a collective. This collective, which inscribes its terms on the individual in her moment of identification, is itself produced through historic processes that include nation formation, the writing of census categories, uneven rights, migration, and systemic exclusions.

At performances by the Beijing Modern Dance Company some audience members self-identified as members of a general category of "Asian," but most wrote a specific nationality: Chinese, Chinese American, Vietnamese, Korean, and Taiwanese (self-expressed identifications also included Jewish, white, Hispanic, and European descent). Identifications that would fall under a census category of "Hispanic" might include—as they did for an evening of son jarocho, an Afro-Mexican genre from Veracruz, in 2002—Chicana, Hispanic, Hispanic (Puerto Rican), half Spanish and Mexican, Latin, Latina, Latino (Mexican Colombian), Latino Peruvian/Mexican, Mexican, Mexican American, Nahuatlaco-Mexica, Spanish, and Veracruzano. Such identifications are at times verbally negotiated. A mother and daughter discussed their ethnicity while filling out the mailing list forms. The daughter told her mother, "Write Hispanic." She herself had written "Spanic." They discussed the issue heatedly in hushed tones, the daughter finally saying, "Okay, I can cross it out." She did not. The mother had written Latina on her form. Later, a young woman said to her friend, "Write Mexican American, because you're Mexican and American." Self-identifications that would fit current census categorization as White included white, Caucasian, German, European Norwegian, Russian Jew, white as can be, white and pink, and American (Welsh, Irish, English, Spanish, French). People also used this space to challenge the question, providing answers such as pinky-tan, spiritual, forced American, human being, citizen of the world!, universal, decline, that is not relevant, and it's offensive to ask that!

The request to "describe your ethnicity" is included with other questions on the mailing list form that provide means of registering membership along different axes, including Grand Performances' audience member ("Is this your first introduction to GP?"), age ("Please indicate your age range"), media use ("How did you hear about this performance or GP?"), and taste ("What types of performances are you interested in?"). Presumed through their inscription on the mailing list form and reproduced by the self-representation of audience members, such categories of affiliation suggest the possibility of knowing and enumerating both the reach of marketing efforts and the composition of the concert audience. Yet this enumeration is always incomplete, insofar as those who report are self-selecting audience members who are not already on the mailing list, who walk over to the information booth to fill out a form, and who usually fill it out only partially, leaving the accounting of audiences undone, to be completed—or not—through other modes of knowing and imagining.

Implicating myself in the regime of governmentality articulated by demographic categorization, while working in the office conducting ethnographic research I quantified the self-reported identities from a season of mailing list sign-up sheets, organizing them roughly by Grand Performances' existing categories. These categories are those of the constituents recognized as already part of Grand Performances' audience, who are programmed for because of their perceived presence in the city. The previous year the publicized audience demographics were changed to better reflect those of the city; having seen the audience, staff members decided that it was not actually 20 percent African American and that there were more Asians. Hence the figures were changed to 14 percent African American and 13 percent Asian, while the figures of 30 percent Caucasian and 30 percent Hispanic remained (see table 1). When my calculations showed only small differences between staff estimations and the demographics self-reported on mailing list sign-up sheets, staff members expressed a sense of validation. "That shows we're not wrong!" one exclaimed. "I think this is pretty impressive," another said. The similarity between the demographics of the audience and those of the City and County of Los Angeles according to the 2000 census was further vindication of Grand Performances' work. They discussed the ways in which the categories might be adjusted: combining the categories of Asian and Pacific Islander would increase that figure, as would doing Filipino programming. The director argued to keep Middle Eastern separate, which affirmed the success of their work with the local immigrant and artistic community.

The categories indicate Grand Performances' understanding and imagining of Los Angeles's diversity, following a trend by which multiculturalism is made in relation to census categories without fully aligning with them. Hence Middle Eastern, which in the census becomes White, was

Table 1. Audience Demographics (percentage)

	GP estimate	Mailing list	City (GP figures)	County (GP figures)
Latino	30	34	46.53	44.53
Caucasian	30	25	29.75	31.09
African American	14	12	11.24	9.78
Asian	13	10	9.99	11.95
Middle Eastern	8	4		
Pacific Islander	2	2	0.16	0.28
Native American	1	1		
Other	2	5		
Mixed		7	5.18	4.94

Source: Grand Performances.
Note: The audience demographics were calculated based on visual estimation and mailing list sign-ups.

given its own category in Grand Performances' estimations. Caucasian was used instead of the census term White, Latino rather than Hispanic. Audience members identifying as "Mixed" suggested the progressive nature of Grand Performances, ahead, its staff said, of local census categories. Acknowledging the high number of people reporting their ethnicity as "Mixed" also reflected the tendency of Grand Performances' staff to shift readily from a fixation on discrete groups to multiculturalism as a whole, ultimately working with people and organizations that, while oriented toward one particular group, would have an inclusive perspective.

Spatializing Grand Performances' demographics, zip codes from mailing list forms showed that the "audience is coming from all over." One staff member said, "This reinforces the idea that downtown is the center of the city and is easy to get to." Grand Performances' own mailing list provided sufficient information to direct marketing efforts toward particular groups through the correlation of geography and identity. In 2002, to advertise performances of modern dance from Mexico City and of son jarocho music from Veracruz, the mailing list was used to find addresses with zip codes in areas known by the marketing director to have many Latino residents. Using maps with neighborhood designations, he determined the zip codes of Montebello, East L.A., South Central, Compton, South Pasadena, and Whittier. In addition, six hundred postcards were distributed through Self Help Graphics, an arts organization in East L.A.

Neighborhood Cultures

One Saturday morning I went to Self Help Graphics with the director of marketing. Self Help Graphics produces prints of Grand Performances' season calendar artwork that are sold at concerts, with profits going to both organizations. The first of these was a painting by Frank Romero, an East Los Angeles native and member of the 1970s Chicano muralist group Los Four. Self Help Graphics is a quick ten-minute drive from downtown, and we took the 1st Street Bridge to cross the concrete bed of the Los Angeles River into Boyle Heights, cutting north to César Chávez Avenue. Traveling west, César Chávez Avenue becomes Sunset Boulevard on the west side of Chinatown, after which it continues on through Echo Park, Hollywood, Pacific Palisades, and Malibu to the Pacific Ocean. We drove east a few blocks to the home of Self Help Graphics' printing studio, gallery, and store, where we were met by the director, Tomás Benitez.

After a tour of the space, he showed us to a table with large portfolios holding examples of their prints; as the director of marketing looked through them for a possible image for next year's poster, Benitez told me about the history of Self Help Graphics. Founded in 1971 by a nun whom Benitez called "Sister Karen," Self Help Graphics began as an art space for youth in East L.A. Its founder believed that art was "a social practice that could build and sustain community" (Self-Help Graphics 2003:196). Benitez first worked with the organization as a teenager, when it was housed in a garage. At the time, he said, Self Help Graphics was focused on Chicano politics, inclusive within the community but excluding others. As director, Benitez was interested in serving the needs of residents of East L.A. while opening the doors of Self Help Graphics to the rest of the city and forging connections with other neighborhood arts organizations. Today its Día de los Muertos festival is attended by thousands of people from across Los Angeles. While the director of marketing and Benitez talked, I looked at the work in the gallery, an exhibit of images and installations of the Virgen de Guadalupe by contemporary artists. The main space filled with an altar, the exhibit extended into the stairwell and upstairs past the gift shop into another smaller room, where small icons, paintings, and drawings hung on the wall.

Grand Performances' longstanding partnerships with cultural centers around the city, including Self Help Graphics, Plaza de la Raza just north of downtown, the Watts Towers, the Japanese American Cultural and Com-

munity Center in Little Tokyo, and the Chinese American Museum near
Chinatown, support its work as a downtown multicultural arts organization
that attracts people from around the city. Grand Performances draws on
the constituencies of these organizations to help constitute its multicultural
audiences. These institutions, while organized around the history and cultural expressions of a particular group, now generally follow the direction
of Self Help Graphics and support a multicultural outlook, working with
other such institutions across the city and inviting visitors from different
backgrounds.

Multicultural arts projects in Los Angeles have a long history as inten-
tional political interventions in the social fabric of the city. Grand Performances draws on this tradition, and the work of the specific organizations
involved in this project in its presenting practices. These projects have been
initiated and operated with the assumption that the arts might help Angelenos come together in harmony, crossing even as they produce ethnic and
racial boundaries. Thus, the urban diversity that has the potential to instigate riots could be deployed as part of a project of cross-cultural understanding through the arts. At the same time, the urban tensions and
violence that multicultural arts projects were intended to help ameliorate
were themselves partial responses to the effects of neoliberal capitalism that
exacerbated already existing discrepancies between rich and poor through,
among other things, the post-Fordist decline of manufacturing jobs in
many parts of the city that resulted in increasingly uneven urban geographies of employment and city services (Kelley 1994; Scott and Soja 1996).

An important predecessor to today's multicultural arts projects is the
now defunct Inner City Cultural Center (ICCC), which also provides a
continuing connection with performance traditions from Los Angeles's historically African American neighborhoods. The ICCC was founded after
the 1965 Watts Riots by C. Bernard ("Jack") Jackson. Jackson, an African
American, was then a dance professor at the University of Southern California–Los Angeles. The ICCC, while providing opportunities for artistic
expression for African American youth, was intentionally multicultural
from the beginning, reflecting actual, rather than perceived, demographics
of the area (Walker 1989). As Jackson explained, "A short walk through the
neighborhood surrounding the Inner City Cultural Center would reveal
that in this community reside people of every conceivable ethnic description" (qtd. in Walker 1989:215). Jackson, in describing his intentions for
the ICCC, emphasized "the artistic responsibility to view art not only as

entertainment, but as a tool to explore problems which exist in our environment" (Walker 1989:92). To this end, Jackson hoped to use the arts "as a tool for helping us solve some of the problems which exist in this society" (qtd. in Walker 1989:49). Multiracial casting for plays reflected the demographics of the surrounding community at the same time as it was intended to show that people of different backgrounds could work together and get along.

The emergence of ethnic-specific arts organizations, many run by people who, like Tomás Benitez, had worked for the ICCC, came with increasing urban segregation. Relying on an understanding of the racialization of urban space that figures the city as composed of marked neighborhoods and a "neutral" downtown, Grand Performances' director explained that the ICCC did not become a mainstream arts organization primarily because of its location in South Los Angeles, which he said was already a racially marked neighborhood feared by some residents of the city. Today, the legacy of the ICCC serves as a means of connecting race and place, supporting a project of bringing otherwise segregated Angelenos downtown through Grand Performances' programming under the name of its founder. For many years, Jackson curated a performance at California Plaza each season, in which he presented jazz artists, the ICCC theater program, and the Watts Prophets, among others. After his death in 1996, artists connected with the center or doing work that maintains its multicultural mission have been presented on a "C. Bernard Jackson Night." The Watts Prophets and Dwight Trible are two of the artists who have performed in this series. Both maintain a connection to Los Angeles's African American neighborhoods. The Watts Prophets, a spoken-word trio credited, along with New York's Lost Poets and Gil Scott-Heron, with creating a precursor to rap, got their start as teenagers at the ICCC. Dwight Trible is a jazz vocalist who performs most often at the World Stage in the Afrocentric arts area around Leimert Park in southwest Los Angeles (see Isoardi 2006).

Cultural centers anchor Grand Performances' marketing efforts in neighborhoods correlated with a majority race or ethnicity. There, other institutions extend these efforts. Grand Performances' marketing for performances of the Beijing Modern Dance Company focused on Chinatown as a neighborhood to locate cultural organizations that would disseminate information to their constituents. The Chinese American Museum helped promote the performances of the Beijing Modern Dance Company by advertising them in its brochure and on its Web site. Bilingual concert programs were distributed at Chinatown service organizations and non-profits, including the Chinatown Branch Library, the Chinese Chamber of Commerce, and the Chinese American Citizen Alliance. Foregrounding the international nature of local multiculturalism, shared background moves from neighborhood to nation, and immigration, a defining feature of globalization, becomes a means of imagining the global city. A translocalization of Chinese politics emerged when one Chinese cultural center would not accept the programs for performers from Beijing because the organization was Taiwanese (Kaplan 1992). Grand Performances' relationship with the Chinese American Museum developed through meetings with its director, as part of conversations about a conceived but never fully implemented marketing project called Tastemakers.

Tastemakers, referring to "people with big mouths," was initiated based on the understanding that 60 percent of Grand Performances' audience members had heard about the event they were attending via word-of-mouth. Grand Performances staff members all knew of audience members who organized groups to attend performances, sent out e-mails about events, and rallied friends and strangers based on their enthusiasm for free concerts. One man who told me he loved attending public concerts explained that he e-mailed a big group of people before each concert; he had celebrated his birthday at a daKAH concert, and had invited a bunch of people, explaining, "It's great because it's free!" Tastemakers was intended to channel these kinds of efforts toward marketing targeted at discrete identity groups.

Tastemakers was an audience development project. Organized around demographics as that which was and could be known about Grand Performances' audience, the project articulated a range of the organization's aspirations. First, concerts would be identified that were understood as appropriate for or of interest to certain demographic groups. Individual Tastemakers would help attract members of these groups to things they would be interested in, with the hope that they would come back to see something else. To ensure marketing would be organized around demographics, Tastemakers would be chosen who were connected to community organizations. As a memo describing the project explained, "They will be selected in relation to specific performances in the summer 2003 season. Tastemakers will be asked to invite people to a performance that they might already attend (i.e. Chinese to Chinese opera) and to another performance they might not attend (i.e. Chinese to Mexican modern dance)" ("Taste-makers 2003," Grand Performances internal memo, 2002).

A letter was drafted to invite select Tastemakers, stating, in part:

We have learned that 60% of Grand Performances audiences heard about us through word-of-mouth. To encourage the growth of this network, we are continuing the Tastemakers Program, begun in 1999. . . .

Tastemakers will help us bring new audiences to concerts at California Plaza. As a Tastemaker, you will be asked to invite people in your community to selected concerts in the 2003 season. . . .

The Tastemaker program helps Grand Performances continue its work in exposing new audiences to the arts and helping connect communities around Los Angeles.

An incentive program based on the number of people who came as a result of an invitation from a Tastemaker would encourage participation and help quantify the results. The possibilities discussed included gift certificates to area restaurants, museum memberships, and free parking for events. As the project was never fully implemented, ideas for how audience members would register their association with a given Tastemaker were not fully developed. After compiling lists from various staff members who articulated their own social networks, four people were approached as part of the Tastemakers project. One was chosen for his access to others who might be Tastemakers, two worked with children, and the fourth was a curator at the Chinese American Museum.

The Tastemakers project revealed ways in which correlations between identity, geography, and taste are made and unmade as presumed essentialisms are questioned and contested. A meeting with the last Tastemaker fleshed out a range of issues at stake. Grand Performances' director explained the project, describing Tastemakers as "people with big mouths." After a short discussion about other organizations already contacted, the Chinese American Museum curator began to unpack the correlations on which the project was based: "In our constituency we have jazz fans, classical fans. . . . It doesn't have to be ethnically specific but place specific. Also, not all are Chinese American, and even if they are, they're not necessarily interested in Chinese things. I came for Ozomatli," a popular East L.A. band. Nevertheless, she suggested the Chinese American Museum could put Grand Performances events on their events calendar and link to its Web site, even choosing particular events as "staff picks." They discussed possibilities for reciprocal marketing, such that Grand Performances' audiences would learn about and perhaps visit the Chinese American Museum just on the other side of the 101 Freeway.

Perhaps, they colluded, a Grand Performances event could be held at El Pueblo, near the Chinese American Museum. The curator from the Chinese American Museum said, "It's the birthplace of L.A. It's always been diverse. The Plaza is beautiful at night, it's lit up and no one is there. They call it 'tourist alley.' There are so many myths and realities of Olvera Street, along with positive experiences." Though "Christine Sterling"—who created Olvera Street in the image of a fictionalized nostalgia for Spanish Los Angeles—"can be called a cultural imperialist, families have valid experiences there." Grand Performances' director concurred: "I was brought down there by my parents to learn about the Mexican history of L.A." Their conversation veered into the history of the area around Union Square, Chinatown, and El Pueblo, where fantasy merged with the lived experiences of Los Angeles's minority groups, producing spaces that continue to support a mix of tourism, boosterism, nostalgia, and everyday life.

> *CAM:* The building that the Chinese American Museum is in was first lived in by Chinese.
> *GP:* The palm trees you see at the exit to Union Station were designed as a marketing tool. Everyone coming from the Midwest would see them and want to move to L.A.
> *CAM:* There were forced evictions. Handbills were posted in English one week before, and few people spoke English.
> *GP:* I remember a kid's history book about Bidwell, who met Sutter, who then had property in Chico. There was anti-Chinese sentiment in that area, but he refused to buckle under pressure and allowed Chinese and Indians to work there. Kearny Street is now the main street in San Francisco's Chinatown.

A meeting about a word-of-mouth marketing project circled back to a discussion of Los Angeles's ethnicized geography. Statements were connected through geography or ethnicity, creating the two as on a similar order in order to move between adjacent buildings and from Los Angeles's Chinatown to San Francisco's. Their exchange articulated key ways in which

Angeleno multiculturalism is produced and negotiated. Little-known histories were drawn together with histories known to be fictional as a basis for celebrating contemporary diversity like that found in Grand Performances' audiences.

The geography discussed illuminates the complexity of correlating place and identity. The historic Plaza just north of the 101 Freeway dates to Los Angeles's founding as a Spanish mission (Estrada 2008; Parson 1993; Ryan 2006). Today largely used by Mexican Americans, it is also the location of the Chinese American Museum. Olvera Street, adjacent to the Plaza, is a short block of souvenir shops and Mexican restaurants visited by tourists and Mexican American Angelenos alike. Items for sale range from Mexican dresses and cheap pottery to T-shirts proclaiming Chicano/a pride. Los Angeles's Chinatown is north of the Plaza. Union Station, across Alameda Street to the east, was built on the site of the original Chinatown, displacing its residents through forced evictions. The new Chinatown is decorated with props from the set for *The Good Earth*. The interplay between cultural space and theme park presented by the area is widely known yet does not prevent Cinco de Mayo events from being held in the Plaza, Chinese retirees from playing mahjong in a Chinatown courtyard surrounded by art galleries, or Silver Lake hipsters from joining Chinese American families for dim sum on Sunday mornings. These kinds of juxtapositions endear the spaces to Angelenos; they appear normal in a city marked everywhere by the film industry and facilitate an openness to people from different neighborhoods. Such spaces are evoked by Grand Performances' staff as quintessentially Angeleno and as reflecting the kind of diversity their concerts are intended to create. As the director explained, Grand Performances' audiences are "multigenerational and diverse. People who want to live in a multicultural community get to have it here. Grand Performances confirms you can live in a great municipality with diverse peoples who are equal." Acknowledging historical injustices and forced evictions relegates inequality to the past, overcome now in a city where other cultures can be consumed equally by all.

While the Chinese American Museum's connection to a Chinese community was the motivation for asking its curator to be a Tastemaker, the museum's openness to others interested in learning about another culture was ultimately what made it a suitable collaborator with Grand Performances. Word-of-mouth is a messy, uncontrollable process that relies on one person telling another, as sounds from mouth to ear create uneven and unknowable networks of knowledge about Grand Performances across the city. Thus the primary way in which most new audience members end up at Grand Performances' events is quickly channeled into an institutional relationship, in which more tangible, quantifiable ways of marketing that can be controlled and known from the side of distribution are used. The Tastemaker's role as a person who comes in contact with others was subsumed by her position as a staff member at a local cultural center. There, Grand Performances' events could be advertised to ethnic constituents on the Internet and with brochures, in turn connecting with general print marketing strategies directed toward particular ethnic groups that include brochure distribution organized by geography and advertising in ethnic mass media.

Translocal Media Spaces

Part of everyday life and the navigation of urban space, media create symbolically meaningful representational spaces and sociabilities (Lefebvre 1991). Structured by and structuring social life, media imbue one's experience in the city, whether in domestic domains or as one moves through its public spaces (Ginsburg, Abu-Lughod, and Larkin 2002; Schein 2002). "The way people understand who they are and how they belong is never anterior to, indeed is inseparable from, the kinds of media they consume" (Schein 2002:230; see also King and Wood 2001; and Naficy 1993). Representing urban space through the use of media to sound an already imagined geography of identity is a spatial practice, deploying media that imbue life worlds of urban residents as they become attached to wider identifications, social networks, and spatial practices that create a sense of the city. Ethnic media provide means and motivations for urban navigation as radio broadcasters encourage listeners to attend concerts, cultural events, and political rallies (Ong 2003). A mother picks up a season brochure at the public library, along with a brochure for a film series and a bus schedule. An East L.A. resident who had brought her children to hear a famous mariachi musician perform in California Plaza the previous summer finds a postcard advertising an upcoming performance of a modern dance group from Mexico City in her mailbox.

Using ethnic media to market international programming helps figure Los Angeles's diversity as global. Spatialized demographics create a map of the city that invokes national identities and diasporic communities. As

Angelenos who might also be immigrants come downtown for performances by artists from around the world, Los Angeles's status as a global
city is at once revealed and created. The limits of the circulation of ethnic
media—which include geography, language, and interest—facilitate target
marketing: "Preexisting forms and channels of circulation" that organize
publics "enable confidence that the discourse will circulate along a real
path, but they limit the extension of that path" (Warner 2002:106). Such
delimitations are productive, contributing to both ethnic solidarity and
self-identification as a member of a given group. The media publics inter-
sected by Grand Performances' marketing extend the aspirations of openness and inclusion associated with urban public space at the same time as
they reflect the reach of the state and its potential to interpellate its subjects
through multiple techniques of governmentality. The possibility of being
part of the face-to-face public of the audience depends on linking into
another source of information that addresses a wider audience than the
concert itself.

Marketing is a tool of both capital and the state that intersects with the
formation of publics structured by categories of multicultural recognition.
Multiple publics organized around music, taste, or identity help create the
audience as the public of the city at the same time as they have their own
organizing principles that can be discursive or practical, with different
demands of interaction and "publicness." Musical publics depend on
shared listeners, taste publics demand an investment in a particular form
or style, ethnic publics ask for the production and maintenance of an identity that is also shared with others, and the public of the public concert
requires attendance at the event. These publics create spaces for the possibility of different experiences of the public concert that can be more artistic,
more social, more personal, or more multicultural. Individuals can be
members of multiple publics that might be organized around an art form
and an ethnicity, a musical style, or a nationality. Affective sentiments of
belonging are often found in the links between multiple publics of which a
person is part.

Attracting "previously underserved" groups provides a means by which
Grand Performances can exemplify its outreach efforts and its ability to
draw new, untapped markets, reaching more of the population of the city.
These groups have included Central and South Asian communities, Hawai-
ians and Pacific Islanders, Latinos, and Chinese. In 2002, initial contacts
were made with Chinese media sources to advertise the performance of the

Zheng Zhou Qu Opera. These included the *Chinese Daily News*, *Sing Tao Newspaper*, and *Pacific Times*, Southern California Asian television channel KSCI, and KAZN Radio Chinese. These media outlets were sent press releases and flyers stating they "were generated from the offices of the Chinese Cultural Affairs Department/Chinese Culture Art Improvement Association." A full-page ad was placed in the *Chinese Daily News* and flyers were distributed in Chinatown and at the Chinese American Museum. The Chinese American Museum's Web site had a link to Grand Performances' site, and KSCI and KAZN mentioned the event on air. These efforts were mostly undertaken close to the day of the performance. Evidence of success in bringing people to the performances was provided by audience members who stated they had heard about the event from Chinese-language media sources, even if this information was conveyed through a complaint that there had not been enough coverage and that it had not been advertised far enough in advance of the event.

To advertise the 2003 performances of the Beijing Modern Dance Company, Grand Performances built on the relationships that were made the previous year. The identification of the staff of the Chinese American Museum as a potential Tastemaker emerged from this earlier work. This year full-page advertisements were placed in several local Chinese newspapers. Grand Performances' staff also consulted with local Chinese community organizations, politicians, and media about how best to market the programs. Along with sources that included the *L.A. Weekly*, *Los Angeles Times*, USC *Trojan Daily*, *La Opinión*, and word of mouth, audience members reported on mailing list forms that they had heard about the event on Chinese AM radio stations (KMRB AM1430 and KAZN AM1300), on KSCI, and through the Asia Pacific Arts Web site at UCLA.

Radio announcements on Chinese AM radio stations in Mandarin prompted many Mandarin-only speaking people to call Grand Performances' office to inquire about the performance. With the help of a native speaker, a transliterated cheat sheet was made for staff members answering the phone to be able to say that the performance of the Beijing Modern Dance Company ("Beijing xian de wu") was at 8 P.M. ("bath di em"), parking was $8 ("pa chuh bath quai"), and that no tickets were needed ("bu yau peyau") because it was free ("sui beyan lai"). We practiced saying the words with the correct intonations, but only the staff member with knowledge of Cantonese was able to remember how to pronounce the Anglicized text. The fact that people called Grand Performances speaking only Mandarin suggested that belonging extended from membership in the media publics of the radio stations to a public of Grand Performances' audience.

A map of KSCI's coverage area reflects the tension between the potential openness of mass media and the delimiting channels of access and engagement organized around interest, background, and language (KSCI-TV 2006). The map shows two expanses of coverage, the larger around Los Angeles and Orange counties and the smaller within San Diego County, reaching the cities of San Diego and Escondido. The larger coverage area spans outward from the border of southwest Los Angeles County and Orange County. The map, seemingly neutral, is legible for those who can decode the representation of Asian populations by city names, whether because of personal membership and residence or interest in the Asian market. Spreading over parts of Riverside, San Bernardino, and Ventura counties, cities are listed that have large Asian populations (Monterey Park, Long Beach), as well as others that provide an understanding of area geography (Burbank, Pasadena). The map represents the ephemerality of broadcast media, spread through the air from one antenna to another or as information traveling through cable wires. Potentially accessible anywhere within the area reached by the strength of the broadcast signal, actual practices limit circulation, creating uneven relationships to the media.

Seemingly disparate information interpolates people's multifaceted lives. The commuter listens to her radio as it broadcasts concert announcements, world news, and information about rallies for immigrant rights. Spanish-language radio stations often announce political demonstrations, many of which—such as an annual janitors' march—are held in downtown Los Angeles. After the immigrant rights march of May 1, 2006, mainstream media expressed surprise at the number of people who participated, unaware that it had been discussed widely in Spanish-language media. At Grand Performances' Central Asian Festival in 2002, many audience members of Middle Eastern descent reported that they had heard about the event on KIRN, a Persian radio station whose listeners often refer to it by its broadcast number AM670. This station was later the source of information about INS detentions following a registration process instigated by the newly formed Department of Homeland Security. Iranians rallied in front of the Federal Building to protest the detentions after hearing about them on KIRN, indicating the significance of the station as a news source for and

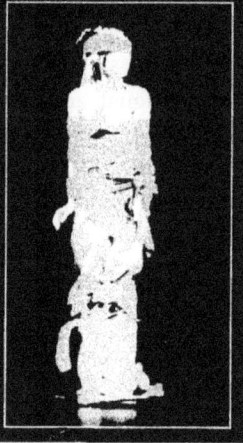

Figure 5. Advertisement placed in Chinese-language newspapers in 2003. Courtesy of Grand Performances.

about the Iranian community as well as a motivator of activities, whether participating in a protest or attending a public concert.

Ethnic media help connect members of a diasporic group, creating new social formations organized around practices at once deterritorialized and reterritorialized (Sreberny 2000:182). Language is a key element in the integration of media in everyday life and its role in sustaining social bonds and identifications. More than half of Los Angeles County residents speak a language other than English at home (Asian Pacific American Legal Center of Southern California 2008:1). Used to support a policy argument for increased English-language education, these figures are only a partial account of Los Angeles as a site for global immigration (Bonacich, Ong, and Cheng 1994; Chinchilla and Hamilton 2001; Der-martirosian 2007; Kohpahl 1998). Studies and census figures cannot do justice to the generations of immigrants who moved to Los Angeles, ongoing movement between countries, everyday realities of immigrant life, the long lines outside the downtown INS office, or the economic relationships crafted with and among industries ranging from restaurants to film.

During a panel at the Latin Alternative Music Conference titled "Reaching Generation Ñ," an editor from *Al Borde* discussed the fact that the publication is in Spanish. Initially, he said to the audience in the Beverly Hills Hilton conference room, the music written about was in Spanish, and readers were Spanish-speaking only. Musicians communicated to the crowd in Spanish, and the crowd identified with that. Now, he continued, things have changed, and readers are 95 percent bilingual. However, as youth lost their Spanish, grandparents became sad, and now people are going back to using the language. Spanish, he said, is becoming the most widely used language in the world among youth. Reading *Al Borde* in Spanish allows young people to get closer to their parents; as one reader told him, "I'm getting closer to my mom. I'm sitting next to her and she reads the paper to me." Panel members emphasized that at the grassroots level in particular language facilitates social bonds around ethnicity and nationality, as publications might be handed out after concerts to a select group known to identify around language.

Los Angeles–area Chinese media target different segments of the Chinese American and Asian population, helping organize the Chinese American population along linguistic divisions that reflect immigration patterns and Chinese politics. KMRB AM1430, once part of KMNY AM1600, broadcasts in Cantonese and Mandarin, filling an "ever-expanding . . . Cantonese market" (KMRB AM1430 2003). KAZN AM1300, on the other hand, was originally K-Asian, which programmed in Korean, Cantonese, Japanese, Mandarin, Polynesian, Tagalog (Filipino), Thai, and Vietnamese (KAZN AM1300 n.d.). In 1993 it shifted to all Chinese programming in Mandarin. Its Web site states, "Unlike other US cities with large Chinese communities, the Los Angeles Chinese market is uniquely positioned as the only domestic Chinese market that is predominately [*sic*] Mandarin-speaking." Yet language is as much, if not more, about China as it is about immigrants to Los Angeles. It continues, explaining, "one must realize and understand that even though many of these listeners may not speak Mandarin as their primary household language, Mandarin is considered to be the 'universal' dialect or language among all Chinese, so, Radio Chinese attracts a broad spectrum of the Chinese demographics" (KAZN AM1300 n.d.; see also Gladney 1994).

The media sources in which Grand Performances advertised appearances by Chinese artists have varying degrees of delimitation. Chinese-language AM radio will very likely only be listened to by Chinese speakers who can understand what is being said. Chinese-language newspapers will only be read by those with reading knowledge of Chinese. While others might hear or see these sources, they will not obtain information from them. Asia Pacific Arts, a Web site hosted by the UCLA Asia Institute, provides information about upcoming cultural events, articles on Asian artists, news from Asian countries, and reviews of past events. Asia includes South, East, and Southeast Asia, and the material is written in English. The Web site is a specialized source that might be known only to those interested in the subjects or issues addressed therein, though any English-speaking Angeleno could ostensibly use it as a source of information for Asian-related topics. Some effort, however, would have to be made in order to find it, either through a search engine, a link from another Web site, or another related source. KSCI, a TV station with pan-Asian programming in fourteen languages (Korean, Mandarin, Cantonese, Japanese, Vietnamese, Arabic, Hindi, Khmer, English, Armenian, Hebrew, Urdu, Tagalog, and Farsi), is the most open and accessible to people outside the ethnic or linguistic publics of these media outlets. While its programming is primarily of interest to those with the national and language backgrounds of the various shows, anyone with a television who is channel surfing will come upon the station and perhaps even pause to watch a show such as one I once came upon that featured Korean women wearing headphones with furry animal ears playing a game of "telephone." Choosing to turn on a channel at a particular time, to watch it or surf past, and the possibility of understanding not only the language but the humor and references are all part of the pathways that help shape media publics.

Language can exclude as well as include. A woman originally from Mexico City wanted to invite some of her East L.A. neighbors—"the cleaning lady, the mechanic"—to concerts of son and mariachi, which she explained was popular Mexican music. Usually, she continued, the Latino programming was more highbrow. She ran a film screening program downtown, had studied economics and worked in film production in Mexico, and was involved in local cultural activities. The fact that the season brochure was printed in English made her neighbors feel the concerts were not for them. Given the high numbers of Spanish-speaking audience members at similar programs, it seems not everyone responds in the same way. Nevertheless, the feeling of exclusion reflects limits of a general audience that is presumed to share a dominant national language. To address such potential exclusions, Grand Performances has developed a Spanish version of its Web site.

Solidarity and belonging structured around identity facilitates marketing, whether for a public concert or a new product. Serving as untapped markets for the expansive logic of capital, identity groups are structured through this process (Dávila 2001). Ethnic media is closely attached to new markets for profit. In using ethnic media to advertise free concerts, Grand Performances taps into this dynamic, deploying market consumers as audience members. This was realized most directly when Grand Performances and KSCI forged a partnership as part of the marketing and sponsorship work for the Beijing Modern Dance Company's performances. KSCI was started by a former employee of Univision; as an Asian, he felt that the Asian American community would benefit from a media company similar to Univision or Telemundo, the major Spanish-language stations in the United States.[2] Asian American is broadly conceived at KSCI; programming includes *Kiss Kiss Kiss* in Korean, *Pin Pon Pan* in Japanese, *Namaste America* in Hindi, *Jaam-E-Jam* in Farsi, and news programs in Mandarin, Vietnamese, Korean, and Japanese. The station is not exclusive to any one Asian nationality but provides a source of affiliation and connection to multiple diasporic communities (Naficy 1993).

Drawing together divergent modes of belonging around identity, KSCI does extensive community outreach, much of which is oriented around the basic needs of recent immigrants. Its director of community relations was interested in doing more with the arts, something, she stated, in which immigrants participated in their home country. Reflecting an understanding of its community outreach subjects as recent immigrants without much money, she said, "Now that they've been here longer" and don't have to worry as much about meeting basic needs "they can start to do those things again." The arts are not only leisure to be consumed after basic needs are met, they also invoke and help produce ethnic consumer publics as part of KSCI's own marketing efforts. In 2002 KSCI sponsored the 1st Annual Lunar New Year Parade & Festival in Hollywood. The event featured a parade with floats and "performers that represented pan-Asian cultures." Live entertainment included traditional cultural performance, rock, hip-hop, and R&B. The parade provided a marketing opportunity that also produced ethnic publics as cultural consumers. According to the KSCI newsletter, "this event demonstrated how providing the right entertainment and attractions can draw thousands of young, frequently elusive target consumer groups" (KSCI-TV 2002:1). KSCI's interest in Grand Performances exchanged two sets of constituents: the needy and the wealthy—those for whom they provided charity and those who would offer new consumer segments for other corporations.[3]

KSCI's donation to Grand Performances was in the form of advertising. A promotional commercial about Grand Performances and the Beijing Modern Dance Company's upcoming appearances was intended to help attract Chinese American—or, more broadly, Asian—audiences to the free concerts in California Plaza. The commercial was aired seventy-six times in July during Cantonese, Mandarin, Japanese, Vietnamese, and Korean programming and ran until the end of Grand Performances' season at the beginning of October. The airtime value for July was $39,000, a figure that allowed equivalencies to be drawn between the advertising of the two organizations. In exchange for the media sponsorship, KSCI's logo was included in Grand Performances' season brochure and the advertisements and programs for the event, and the station was thanked during the director's pre-performance announcements. KSCI benefits from this kind of advertising, in turn banking on its market of Southern Californian Asian Americans to appeal to other companies that might advertise or invest in the station.

Ethnic consumer publics are construed as potential markets for capital that in turn can inspire feelings of belonging on the part of ethnic subjects who feel newly included in the general populace. As KSCI's material asserts, the Asian market is "a young, rapidly growing and affluent consumer mar-

ket. Delivering a marketing message through Asian-language media connects a company to the hearts of Asian-American consumers. In an age
where . . . new market growth is key, Asian consumers are a critical group
that has the elements to increase a company's profits" ("KSCI-TV" 2002).
KSCI's advertisement, printed in *Advertising Age* next to a piece titled
"KSCI-TV Reaches Millions of Asian-Americans," shows a walk sign with
an arrow pointing up and text saying "Asian Market." The blurb next to
the graphic states, under the KSCI-TV logo, "Your direct connection to
Asian American consumers." This consumer segment has concrete economic value: "At 12 million strong, the Asian-American population is just
one-third the size of the Hispanic population but with an annual buying
power estimated at $254 billion" (Wallenstein 2002:1). At the marketing
panel during the Latin Alternative Music Conference, representatives of
companies and Spanish-language music journalists emphasized how inclusion as consumers contributes to a positive sense of recognition (Latin
Alternative Music Conference, Marketing and Branding Panel, "Reaching
Generation Ñ," August 14, 2003). Latino youth, one panelist argued, are
saying, "We have money, we buy stuff and go to concerts," and in this role
they are defining what is going on. Another emphasized that Latino culture
is becoming a worldwide culture, not just a local culture, as artists like
Christina Aguilera and Ricky Martin are recognized as mainstream pop
stars, superseding any particular identity. Similarly, a show on MTV called
Yo MTV Raps that started on MTV Español is now mainstream. The Latino
market, one panelist concluded, is an untapped market.

Conclusion: Marketing Events

A plastic maraca, still in its plastic bag, lies on my desk. The end of the
handle doubles as a bottle opener. The red shaker is graced with a Wells
Fargo logo. Wells Fargo Center, developed by Maguire-Thomas, is on the
other side of Grand Avenue from California Plaza. Its red granite edifice is
graced with modern sculpture by Dubuffet, Miró, Robert Graham, and
Louise Nevelson. This night the bank is a sponsor for an evening of Afro-
Cuban music. Later in the evening, the maracas are taken out of their plastic bags and shaken in time with the beat, as capital, through marketing,
facilitates feelings of groupness and pleasure, allowing audience members
to make the same beat as the musicians and as each other.

Public concerts look to the city for their audiences, turning back on themselves to frame their significance, as performances are both occasions for marketing and marketing events in themselves. Asserting that the concerts are the best advertisement for Grand Performances, journalists, politicians, and donors are invited to events they might especially like, their privileged position acknowledged by the designation of reserved seating for members of these groups. Marking its status as a civic proxy in the neoliberal state, capital and politics are brought together at Grand Performances' events. The information booth is referred to in the office as the "marketing cart." Here, before every performance, brochures for Grand Performances' events are laid out, along with flyers for downtown destinations, other concert series, and upcoming events, as well as mailing list forms and a small red bucket for donations.

Brochures visually represent the experience that supports the recognition of the city at the concerts, with images of music and dancing amid skyscrapers. The brochure created for Grand Performances' fifteenth season featured a painting by Frank Romero, an East Los Angeles native muralist whose poster Self Help Graphics had printed. Buildings recognizable as City Hall, the public library, and the skyscrapers of Wells Fargo Plaza and California Plaza surround the top and left side of a stage on which a man in a sombrero plays accordion, his shadow cast behind him by the yellow spotlight of the evening performance. Angel's Flight Railway, the risers of the Marina Pavilion dotted with audience members, and the canopies of the Plaza surround the right side and bottom of the stage. Behind, stars shine in the night sky.

The following year, Bernard Hoyes, a Jamaican painter based in Southern California, used his signature style to create a slightly more abstract representation of the venue. Brown-skinned women dance in swirls of color, some raising their hands as they move their hips, others beating drums that radiate out into multihued circles. He was asked by Grand Performances to add high rises in the corner to more clearly depict the setting of the concerts. These appear in the right-hand corner, evident as city hall and the canopies of California Plaza. Palm trees evoke the L.A. skyline. The figures convey Grand Performances' aspiration of inclusion, sustained by representational images that carry a message that specialized knowledge is not required in order to enjoy the concerts. Visualizing music, they suggest the possibility of a sensory experience of sound and city.

The city is sounded through media that reach out across the geographic expanse of the Los Angeles region, mapping the metropolis through the recognition of identity categories that serve as existing means of measuring and marking the city's geography. Media share the material and ephemeral nature of sound, volume's dual nature as amplitude and size. Circulating through airwaves, voices of friends, and printed flyers, events are announced that will take place once, requiring movement from one part of town to another in order to hear a performance with others who traveled from around the city for the same event. Temporal as well as spatial, the media of concert marketing make apparent the temporality of space and the spatiality of time. Past and present come together through the flow of media and people across the city, as real and imagined ethnic neighborhoods are connected to longer histories of multiculturalism in Los Angeles, institutional sites, and responses to urban dynamics ranging from tourism to riots.

3

Performing L.A.

Music has an image-repertoire whose function is to
reassure, to constitute the subject, who hears it.
—Roland Barthes

DaKAH's cross-pollination of jazz improvisation, classical
discipline and hip-hop spontaneity is uniquely Angeleno.
—Natalie Nichols, 2004

Double G steps onto the podium and the orchestra tunes to the piano's A minor chord, with D-A-F played sequentially and held to sound a chord. Cellos start. Along with the other cellists, I play my A string, then D, G, and C. At first the sound is that of a symphony orchestra tuning: strings and brass sustain their notes, adjusting to match the pitch of the piano. Then, as the strings finish, a horn plays a jazz riff. And throughout, if you listen carefully, you can hear the DJs softly playing beats with their drum machines. As the orchestra quiets, Double G raises his baton to cue the musicians and lowers it for the beginning of *Gang Starr Remix*. The drum machines, bass, and strings start with the drum kit. "Hold your hands up," the female MC sings. "O-o-oh o-oh oh." The rest of the women singers chime in. "Uh," a male MC interjects.

DaKAH is a hip-hop orchestra of more than sixty pieces, with MCs, DJs, horns, winds, strings, and percussion. Double G's creation in elementary school of an instrument that could play every sound in the orchestra has taken on a prescient quality. With daKAH, Double G makes his instrument anew, bringing it into being with the help of musicians who bring training and experience in different musical styles to create a hip-hop orchestra. The name daKAH, which Double G told a reporter "has roots in ancient Ghana," provides a social model for this instrument; it refers to "a policy of societal organization and an understanding of the creative process" in which people are given tasks according to their skills, thereby creating "a feeling of meaningful work amongst people, and the satisfaction in feeling like you're on your right path" (Oshun n.d.). He started capitalizing the second syllable to assist with pronunciation. DaKAH mixes up the European symphony orchestra, taking it apart and putting it back together with sounds from popular music.

Grand Performances' program notes for the evening emphasized the diversity of musical influences that come together in daKAH: "Charles Mingus, Duke Ellington, and the spiritual rhythms of Cuba, India and Africa" are "entwined" with genres and instruments that include "European classical music, improvisational jazz techniques, singers, rappers, brass and woodwinds, violins, violas, drums and bass, guitars and harps, turntables, congas, rhythm and rhyme." Of Los Angeles, "All converge to create some of the most progressive and fresh new music being made anywhere in our city." A notion of mixture, of the convergence of musical styles and traditions, is congruent with Grand Performances' programming and its aspiration to "reflect the diversity of the city." Adjectives of "fusion" and "eclectic" used to describe the music figure the performers in and of a Los Angeles defined by its diversity, at once evoking and creating, imagining and performing, this city. Evocative parallels rather than metaphor for each other or for the city, musical mixture and social mixture are two coexisting but distinct frames that align in particular ways as correlations between art and the social, music and metropolis, are presumed and produced.

At daKAH's California Plaza performance, the possibility for relationships to be drawn between performance and the city transpired primarily in the composition and performance of musical mixture, in the formation and delineation of genres as supporting intergenerational understanding, and in lyrical representation. Grand Performances' project of facilitating cross-cultural understanding lay within this overarching frame. DaKAH creates a space for a coalescence of social imaginings insofar as it falls within a definable genre of hip-hop that came out of the context of the American inner city. It is read as European classical through the presence of orchestral instruments and elements of the musical composition. Organized around generation, hip-hop and orchestra are disarticulated in order to map the

former onto youth and the latter onto Grand Performances' "older, established audience." Yet the blending and crossover that happen through composition and performance create something musically that evades and
destabilizes a simple social mapping of music and identity. Hip-hop and
orchestra are both made and unmade in daKAH, at once supporting and
undoing presumed social significations. While hip-hop and orchestra are
recognizable forms that are stable enough to invoke correlations with generation, these genres are constantly negotiated, such that hip-hop becomes
hip-hop *orchestra* in composition, and orchestra becomes *hip-hop* orchestra
in performance. Nevertheless, daKAH's ability to signify a diverse city is
not undermined; rather, intertextual gaps opened by the undoing of genre
are re-closed in the service of pervasive interpretations linking Grand Performances' concerts to urban diversity. As such, daKAH can become "mixture," invoking a diverse L.A. imagined as the sum of its parts.

Sonic Embodiment

DaKAH is first and foremost a hip-hop group. Yet as such, daKAH reflects
how hip-hop—and genre more generally—is constituted through negotiations over musical and social boundaries. Along with original composi-
tions, daKAH performs arrangements of existing hip-hop recordings and
other work that has been influential for hip-hop. Artists whose music Double G has transcribed and orchestrated for daKAH include Gang Starr, the
Roots, and Parliament Funkadelic. DaKAH can be categorized as part of a
trajectory of hip-hop that includes both a West Coast tradition of drawing
on funk music and what Krims categorizes as jazz/bohemian hip-hop
(2000:65; many of these groups are also defined as West Coast underground
hip-hop). Artists in this category include A Tribe Called Quest, the Black
Eyed Peas (with whom a member of daKAH plays), and the Roots. While
the fixity such categorization suggests is debatable, its significance for
understanding daKAH as a hip-hop group is that, following Krims, this
category of hip-hop attracts audiences with a range of racial and ethnic
backgrounds. Moreover, as suggested by the fact that these groups are from
across the United States, jazz/bohemian hip-hop is not measured by the
East Coast–West Coast rivalry of the early 1990s. Reflecting its range of
musical influences, daKAH got its start playing in a club in Santa Monica
that primarily showcases DJs and world music. Members of the orchestra
play and sing at open mic poetry and musical jam events, as well as in

studio sessions, jazz ensembles, hip-hop groups, rock bands, and symphony orchestras around the region.

As a hip-hop orchestra, daKAH is pressured to conform to definitive genre features. Accusations of not being "real hip-hop" begrudge the hip-hop status of daKAH while pointing to wider processes of defining hip-hop and musical genre. Genre is both musical and social, constituted through inter- and intramusical identifications.[1] At daKAH's California Plaza performances, genre was constituted through a combination of elements including instrumentation, composition, aesthetics, performance style, audience response, taste, and venue. Albeit flexible and open to multiple meanings, genre carries with it certain defining associations and depends on a level of consensus and consistency in interpretation. The boundaries of hip-hop are drawn around discrete and overlapping central categories of race, geography, instrumentation, economics, and lyrical content (Forman and Neal 2004; Light 2004; Samuels 2004). Boundary-marking evaluations focus on particular moments of a complex social and musical trajectory of hip-hop, fixing certain contributions and elements as defining features of the genre.

Social correlations are a dominant framework for defining hip-hop. Of these, race in particular is central to marking boundaries of the genre, and hip-hop's African Americanness is negotiated through authorizing histories, lyrics, and consumption. Although hip-hop has been widely described as an African American art form because of its roots in 1970s Bronx and the social concerns of its early participants (Rose 1994), its mixture of Puerto Rican, Jamaican, and African American contributions has recently been written into its history (Flores 2004). In hip-hop lyrics "blackness" is invoked through markers that include neighborhood, national geography, and food. These are in flux, and one might take precedence over another. Double G addresses the overlapping identifications that emerge in moments of genre formation; shifting from race to the city, he asserts, "Hip-hop is an urban American thing. A white kid from the Bronx is gonna understand a lot more about hip-hop than a black kid from Boise." In his experience, complexities of race and place provide a metaphor for criticisms about daKAH's hip-hop status: "Just like in the hood, a kid that is light-skinned gets fucked with by everyone. He's too light for the black ones, and too dark for the white ones."

Double G identifies himself according to social practices, neighborhoods, and musical experience before ethnicity:

This is the demographic I represent. I've spent a large chunk of my life living in working-class neighborhoods, both in the suburbs and the inner city. I've spent a lot of time on the road traveling with musical ensembles, which has broadened the approach I take with different people. My ethnic background is Mexico/Italy/Ireland/Spain/Native. Basically, America before America was USA in some corners.

He keeps his hair in two long braids, practices martial arts with a machete, and is often seen wearing an L.A. Kings or Oakland Raiders jersey. Double G situates his identity in relation to specific experiences that have given him commonalities with others who share those experiences. In not announcing or projecting a sense of a reified "identity," Double G lives Frith's call for an understanding of identity as "*mobile*, a process not a thing, a becoming not a being" (1996:109). Music is a key element in one's identity formation, and "our experience of music—of music making and music listening—is best understood as an experience of this *self-in-process*" (Frith 1996:109). The self-in-process Double G describes as emerging from music is not a direct mapping of ethnicity and genre but an increasing openness to difference. This, too, can be a form of identification, of situating oneself in the world.

Venue, important for marking and making genre, is also an element whose significance can be undone. At California Plaza, the venue was said to "elevate" hip-hop to the level of a formal concert. A horn player in the group stated that California Plaza was daKAH's best venue and that a performance there was a "concert" not a "gig," distinguishing a concert as a formalized event with a structured time frame and audience practice and a gig as a show in a club that might start late, not pay, and have an inattentive audience. Double G credits Grand Performances with making "it possible to bring our orchestra out of the clubs in Los Angeles and into a concert setting. The beautiful venue attracts an audience that was previously unaware of what we are doing." At the same time, venues are malleable, and their genre signification can be undermined or altered through the music performed. California Plaza, with its outdoor amphitheater that is a fountain during the day and a stage at night, is not clearly definable in terms of genres, especially since many different musical forms are performed there. Moreover, daKAH transforms the space. At daKAH's performances, California Plaza feels like a club, as the energy of the audience overcomes the rigidity of the corporate plaza.

As both hip-hop and orchestra, daKAH negotiates the possibilities of formal musical aspects of hip-hop such as instrumentation and composition. Although the first mass-market hip-hop recording, Sugar Hill Records' "Rapper's Delight," released in 1979, incorporated live instrumentation, its use in hip-hop is still contentious (Greenberg 1999). Schloss found, in interviewing hip-hop producers, that "the use of live instrumentation is considered legitimate by producers only when three conditions are met: when the live musician understands (or at least capitulates to) a putative 'hip-hop aesthetic,' when the instruments are used to support musical themes that are already apparent in samples, and when they have the 'right' timbre or ambience" (2004:69). Other such boundary markers include the difficulties in allowing hip-hop to be both electronic and live, to be simultaneously party music and to express political concerns, to be truth and exaggeration, and to be marketable and to reflect sometimes shocking or offensive positions from outside the economic mainstream.

Attempts to definitively define hip-hop are made on axes such as these, even when both sides and more might be true simultaneously. Gangsta rap is described alternately as a reflection of an urban reality, an exaggeration, and a means of making a lot of money in a neoliberal consumer economy that allows for the massive consolidation of wealth by select individuals (Comaroff and Comaroff 2000). The opportunity to make money as a hip-hop artist (or other kind of mediated entertainer) is close to the surface in Los Angeles, home to many of the music industry's record labels, recording studios, and video production houses, as well as individuals who provide examples of having "made it" in the entertainment industry (Shaw 2002). One of daKAH's MCs addressed some of these tensions in asserting that daKAH should be considered real hip-hop:

> "People were complaining about the materialism in hip-hop and the lack of live organic music in Los Angeles. A lot of musicians were really hungry to do something that meant a little bit more than how many people you can draw," vocalist Juliana Jai told the *Los Angeles Times*. "What we're doing is way more hip-hop than what any keep-it-real kid with a drum machine can do." (Russell n.d.)

Jai recuperates hip-hop as anti-commercial, as a practice that has meaning around musical production rather than the marketplace, and as expressed

through the live performance and instrumentation of daKAH. For Double
G, making money through corporate sponsorships, session gigs, record
sales, and a potential record contract is not incompatible with playing for
the love of the music, being creative, and expressing progressive politics.[2]
In being "way more hip-hop" by playing traditional instruments live,
daKAH also helps support Grand Performances' project of intergenerational understanding that entails highlighting hip-hop and orchestra as distinct elements that correspond, respectively, to a new youth audience and
an existing older audience.

The codes, aesthetics, and distinctions that constitute hip-hop as a
musical form provide the foundation for Double G's compositional practice. Through composition a hip-hop audience is addressed, invoked, and
constituted through inclusions and exclusions organized around a particular musical competence. Following Bourdieu, "A work of art has meaning
and interest only for someone who possesses the cultural competence, that
is, the code, into which it is encoded" (1984:2). Applicable more broadly
than as a basis for class distinction, for daKAH this can mean that while
there are those who denigrate the music as "simplistic" or not following
certain compositional rules, there are also those who have the cultural competence to understand and appreciate the music. Double G's compositional
practices are legible to competent listeners who understand and respect
what they hear. Thus, while a horn player attending another California
Plaza concert criticized daKAH's composition, stating that he did not like
all the parallel fifths (something European classical music theory classes
teach is against the rules), another musician, listening to the recording of
the daKAH performance, commented, "I like the open fourths." More
clearly exemplifying the possibility of having an insider status with regard
to hip-hop and daKAH, a musician friend listening to the CD noticed that
the orchestration was an instrumental arrangement of the drum machines
from the original track. Double G, on hearing this, exclaimed, "Someone
who was a hip-hop fan? He picked it out? He heard what we were trying to
do? Yes! That's like the best compliment right there. That's like the biggest
thing right there. That's it."

The harmonic material a hip-hop audience appreciates in the scoring is
complicated in Double G's composition process, as a use of technology that
is assumed to provide legitimacy for one genre is transformed into live
instrumentation. Technology is an undeniable fixture in musical practices,
foremost through recording technologies that provide the primary mode

through which music is consumed. Hip-hop was central in developing the use of sound technology as musical instrument, especially in its incorporation of record players, samplers, and mixers (Katz 2004), instruments that now define the genre. In moving from recorded music to a handwritten score that will be performed live, Double G undoes the technologization of music while he taps into the pervasive presence of technology in musical experience.

The California Plaza concert began with *Gang Starr Remix*. New York–based Gang Starr is described as bringing jazz into hip-hop with socially conscious lyrics. Its musical intentions draw explicitly on an African American musical history. In the liner notes to *Step in the Arena*, the members of Gang Starr end their thank-yous with "peace to the Nation of Islam, 5 Percent Nation and all religions of the world that uplift the black race and to everyone who supports real Rap Music" (Gang Starr 1991). Double G had the idea for *Gang Starr Remix* when he was at a party and "Discipline" came on. He had always liked Gang Starr. Hearing the song in a "cool, really fun party atmosphere," he had the "desire to do Gang Starr." After choosing a few sections to loop from "Discipline," Double G chose eight other Gang Starr songs he liked from his iTunes collection. These were "Ex Girl to the Next Girl," "Words I Manifest," "Code of the Streets," "Mass Appeal," "Soliloquy of Chaos," "Royalty," "Betrayal," and "Jazz Thing." Double G then imported the songs from iTunes into ProTools. In ProTools he took eight-bar sections that he liked or thought would be good for daKAH to play. He looped these sections and made separate audio tracks of each, which he called "score pages." He then arranged the short sections in a row that would correspond to the final composition. He listened to the loops over and over, transcribing them and orchestrating them for daKAH's instrumentation. He wrote the score by hand on staff paper and a member of the orchestra prepared the individual parts.

During this process, through listening, Double G started "breaking down what the loop is actually made of." After organizing the melodic and harmonic content, he made decisions about which instrument should play which part. From there, he took creative liberties, varying the original: "Like when it goes into a minor part of the song, that wasn't originally part of the track, that was just me going, okay, by about now I myself am bored of hearing this shit so I want to hear some different shit but I want to stay with 'Discipline,' I want to stay with the same song, so we'll just change the harmony up a little bit inside." While repetition is an element of nearly all musical genres, the loop is a defining feature of hip-hop, emerging with the deployment of musical technology for music making. As Schloss writes, "hip-hop in general—and the sampled loop in particular—is a logic of musical repetition *as* artistic differentiation; the producer's creativity lies in the ability to harness repetition itself" (2004:138). Double G combines this hip-hop approach with approaches he learned as a composition student at Boston's Berklee College of Music and from playing saxophone with jazz, funk, and classical groups, incorporating a range of musical influences that lie outside those that are relevant for Grand Performances' aspirations of intergenerational understanding.

Drawing on multiple musical worlds to create a singular sonic experience, Double G explained how he composes with two harmonic aesthetics in mind: the sound of two turntables played by a DJ and Mingus's notion of "strong melody":

> The whole idea of hip-hop is, you've got a record here and a record there, and so, if you mix this beat against this beat you're going to end up with notes that theoretically clash, you know, and that's what gives it the sound that it has. A lot of the harmonic decisions come from trying to make it sound like we're generating music from two different records. So the brass and the woodwinds will be in one key and the strings will be in another key. If I turned it in for my theory class I would have gotten an F. The idea is you've got a song over here that sounds really good and complete by itself, and so in coming up with the two different sections, they have to stand on their own, so if I pull the strings out, the brass are still playing and it sounds like a full melody.

Mingus, he said, "taught, as long as you have a strong melody against a strong melody, you can have clashing all over the place." Thus, when Double G changes the harmony in *Gang Starr Remix*, the vocalist can keep singing the original part: "With the Mingus theory of the strong melody against the strong, melody will survive. So you've got the singer, that when they sing over the top of that and you play it on the keyboard, there's no reason it should work ever. But if you just listen to it with the beat, the beat's strong, the singer's strong, it's been established at the beginning, so everybody's ears are like, 'That's the hook, yeah, that's the hook,' but they don't realize the orchestra's playing a completely different key right now."

Jazz, already integrated into hip-hop by artists such as Gang Starr, is deployed to create "the hook," which brings listeners into the presence and present of the musical performance.

The "hook" is constituted for and by the audience, who feel drawn into the music by having a recognizable, repeating part that is catchy and easy to sing along with. Double G assumes audiences will listen primarily to this part of the music, not the harmonies. Within each of those two parts, however, the singer's and the orchestra's, the harmonic arrangement is relatively standard and "to the book," adhering to chord progression "rules" of music theory and composition in a popular music tradition. Several sections of the orchestra will have the same rhythm and will play the part in harmonies that consist of simple chord progressions. In playing these parts, the musicians embody the musical history created with the composition, producing genre through physical practices.

"Code of the Streets," the fourth song in *Gang Starr Remix*, starts with the strings. The violins play sustained half-note chords that descend by half-steps. The cellos play a part that is a combination of the drum beat and the bass line of the original. The physicality of playing an acoustic instrument and the musical technology remain intertwined through processes of learning and performing the parts. The drummer, who also started at the top, brings the bass drum in at the fifth bar. A DJ scratches a record, making a high sound that comes in every few bars. In the cello part, two bars repeat, but the fourth time the pattern changes and there is a rest where the first time the note was held.

We practiced the part in rehearsals. Double G played the loop of the original recording for us. He emphasized the rest, telling us how the note should have as clear an end as a beginning. We were supposed to match the sound of the MPC. The Akai MPC 2000 has been one of the most desirable sampler/sequencers for hip-hop producers. It was a coveted item among the aspiring L.A. hip-hop artists interviewed by Shaw in *Westside* (2002). Schloss (2004) describes how the emergence of samplers allowed hip-hop artists to do what they had been doing with records and turntables in a more controlled and extended manner. Now breaks and other material could be sampled and sequenced, reworked and combined with other material to create beats and hip-hop tracks. The MPC has a specific sound and, like other samplers, repeats the same material identically every time. As Double G explained, the MPC sampler "just generates a sample, so no matter how hard you hit the pad it's the same exact attack every time. By

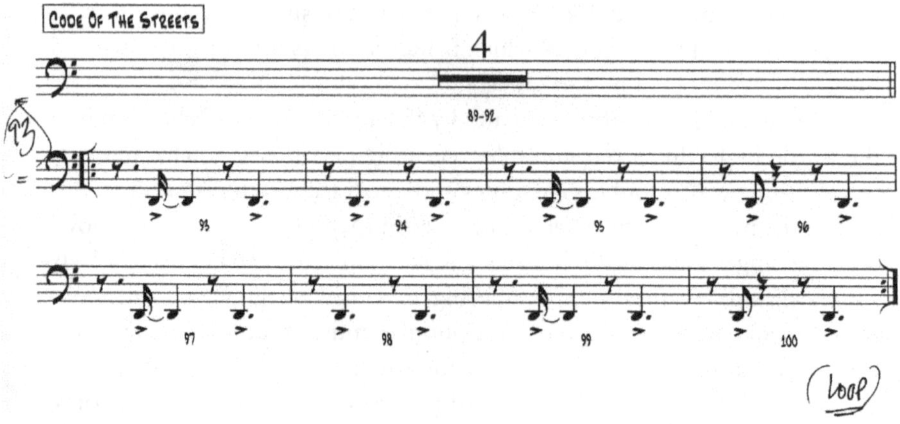

Figure 6. Cello part, showing the third and fourth repetitions of the pattern. Note the dotted quarter note in the second bar that in the fourth is an eighth note with quarter rest. From "Code of the Streets," by Guru/DJ Premier, arranged by Stravingus. Courtesy of Geoff Gallegos.

the same token, it has the exact same release because it's the duration of the sound. You're basically trying to play the exact same thing every time." This repetition and a particular sound he characterized as "hka, hka, hka" are integral to hip-hop tracks. Translating this sound to live performance takes discipline, he said. But "it's a different kind of discipline. It's not a discipline of technical virtuosity or anything like that. It's almost a more tai chi approach to music. You have to have the same exact attack every single time." Playing the sound of the drum machine on acoustic instruments requires a particular frame of mind and an understanding of what he is aiming for: "It's real, if you're not into it with that mind-set, it's real fucking boring, 'cause it's the same shit over and over again, but once people understand what I'm trying to do, people are like, oh, that's what he's trying to do."

We repeated the part over and over, trying to match the sound of the MPC we heard on the recording. This meant using no vibrato and working with our bow arm to articulate the beginning, middle, and end of the notes in a uniform way. There should be no audible attack at the beginning, no decay through the note, and a clear cutoff of sound at the end. Our right arms had to stay active and engaged, held more rigidly than usual. We were trying to do what Schloss argues is impossible, "for those who have been educated in this aesthetic, a beat created with live instrumentation alone can lack a sense of purpose, urgency, and direction; it has no center. At the same time, it also lacks boundaries" (2004:69). Working on the sound production alone was not enough to create the correct sound. Double G was unhappy with the fact that our instruments rang during the rest. "Damp the sound," he said. So we put our hands over the string we were playing, which were the G and C strings, respectively, for the first and second cello parts. They still rang. It was our D strings, which, open, reso-nated with the D from whichever string we were playing it on. So we cov-ered all the strings with our fingers together, making a paw of our left hand. This worked, and we moved on. Later, in rehearsals and performances, when Double G wanted to remind us to play like an MPC, he would jab his finger at an imaginary waist-high drum machine, each jab indicating an identical, rhythmic impact to the one before and after.

Through the process of learning to make a cello sound like a drum machine we were engaging in and producing musical meaning. As classi-cally trained cellists attempting to replicate the sound of a sampler, we were embodying another musical world with a distinct aesthetics, politics, and history. Music is always embodied: the phrase that tears at your heart is created by a physical gesture or a pause in a vocalization. In learning "tech-nique" musicians are also embodying a musical history with particular aes-thetics of phrasing and articulation. When there is an apparent seamlessness between the instrument and the genre the phrasing appears "natural," somehow unworked or unplanned. In "crossing over" and play-ing a genre with an instrument from another tradition, the work of musi-cality can become apparent in that disjuncture. As string players playing drum beats, we helped situate daKAH as a hip-hop orchestra in which both aspects were present, with hip-hop found in the beat and orchestra in the instrument.

Instruments have residues of social histories that contribute to their perceived associations (Qureshi 2000). Yet as with musical genre, instru-ments can mean different things in different contexts. In the case of daKAH, sedimented meanings associated with the physicality of the Euro-pean classical instrument and the sound of those instruments carry across genre, giving a sociomusical status to the ensemble that does not depend on the musical material played on those instruments. Audience responses to daKAH indicate how musical genres are made through both aural and visual recognition. Seeing and hearing strings, bassoon, and pedal harp help

legitimize the ensemble for those who might be turned off by a group playing only electronic instruments or turntables. For daKAH, this is the means
by which Grand Performances' desired older—or classical—audience is
drawn to the performance, helping mitigate criticisms of hip-hop that focus
on the lack of traditional instruments: "Those who are critical of hip-hop
have never been shy about equating the lack of live instrumentation with a
lack of musical quality. . . . Such critics seem to operate from a sense that
the use of live instrumentation is self-evidently superior to the use of samples" (Schloss 2004:76).

Having European classical orchestral instruments onstage invokes associations of taste, musical legitimacy, and, in particular, virtuosity, which
has become a defining feature of live performance and an evaluatory measure in the context of electronic technology. An audience member articulated a heightened version of this perspective, saying, "There's something
about seeing people do physical things that other people marvel at, and
they'll pay to watch people do physical things. It's a wonder to watch a
great musician play, to physically execute their skill, to move their fingers
across the keyboard, to move the bow across the strings of a violin or finger
a clarinet." Seeing hip-hop played on European classical instruments reinforces the effect of virtuosity that has taken on particular significance for
evaluating music from the Romantic era onward, with musicians and composers increasing speed and technical difficulty. Now, in order to play the
repertoire at the level of the standards that have emerged in part from a
century of recorded music, people begin playing orchestral instruments
when they are very young and spend many hours practicing to be able to
play well. This labor is seemingly conveyed by the sound of the instruments,
frequently described as beautiful and expressive.

With daKAH, the visual recognition of the instruments invokes this
meaning. Aurally, the expected sound of the instrument is undone in its
unexpected parts. In daKAH, strings, winds, and horns most often play a
repeated part that follows the aesthetic of the drum machine and the loop:
play the same thing exactly the same way each time. The DJs and MCs are
the ones directed in the score to display virtuosic skills, the DJs on their
turntables and the MCs with words, in the form of good rhymes, a good
beat, and sometimes, if they choose to do so, speed. Neither of these dis-
plays of virtuosity is visible to audience members. There are no fingers
moving fast, no bows blurred by violent string crossings. Instead, the turntables are placed flat on the table, and the only visible movement is of the

DJ pressing his head to one side of the headphones; his virtuosity—like that of the MCs—is aural. This musical crossover disrupts a simple mapping of social group and genre, complicating the Bourdieuian taste categories on which generational musical-social correlations depend.

Both of these core elements of hip-hop require long hours of practice to achieve a level of mastery. Yet instead of learning from teachers in a traditional master-student relationship like that of European classical music, DJs and MCs generally study other performers, both live and on video. A California Plaza security guard became interested in hip-hop after seeing a hip-hop dance performance at California Plaza a few years before. Seeing *Scratch* subsequently inspired him to become a DJ. He bought videos about the history of hip-hop, and when he watched *Scratch* again he knew who the people were. He used his skills to make a little extra money playing at parties and weddings, mostly salsa and Top 40; "what people ask for," he said, though he had also taught himself to scratch. Like classical musicians, DJs and MCs practice the techniques alone, at home, in preparation for potential public events. The history of hip-hop is one of a drive for constant virtuosic improvement on what came before. MC and DJ battles have been key means of fostering this tradition, in which the audience becomes the vocal judge of dexterity, creativity, and virtuosity of one MC or DJ against another.

Hip-hop and orchestra merge in the ways in which daKAH plays together as an ensemble. A kit drummer provides the primary beat for the orchestra. He starts the piece and the cellists play the bass line, the part of the music that feels like it beats in your chest. We dance in our seats as we play, and when we rest we bob our heads along with the continuing beat. We feel the beat more than see it, as it comes primarily from the drummer rather than the conductor. The energy is high, as the beat played by the bass and percussion maintains a good dancing tempo throughout every piece. A cellist said that playing in daKAH is "fun. It's easy and you just dance," a marked distinction from what cellists do when playing European classical music, where they sit still or sway expressively with the melodic phrase. The opportunity to "just dance" is a powerful motivation for playing in daKAH; nearly everyone in the orchestra performs every concert, paid or unpaid. Double G asserts that daKAH's longevity and success rest on a good sound, the participation of the musicians, and fun: "As long as I think the ensemble sounds good, and as long as the cats keep showing up to play and everyone is having a good time, we have succeeded."

One of the DJs also plays an MPC. The rest of the musicians follow this auditory beat to keep time within each measure. Unlike a classical orchestra conductor who directs the entire piece, Double G gives cues for beginnings and ends of songs, as well as when to stop playing one section and move to the next if the section has been looped for longer than is indicated in the score. To cue the orchestra he holds up a notebook with the measure number of the section about to be played, and uses his fingers to show the number of measures remaining until that measure number. One beat before the section he gives an upbeat, directing the subsequent downbeat toward the section playing the main part. As the conductor, Double G is responsible for listening to the whole group and determining the shape of the music. The members of the orchestra listen to the other musicians, working to play together in order to realize the score and create the music. When individual musicians take solos they orient their playing toward the audience. Double G interpolates between the musicians and the MCs, keeping the whole orchestra together. He will extend a section if the MCs are going strong and "something is happening" musically. At this point the other parts "loop" the section, providing a steady foundation for the creative work of the MCs. The MCs are the interface with the audience, directing their lyrics and presentation outward.

Civic Harmonies

Organizing a moral community of a diverse city, public concerts also aim to transform audience members. With daKAH, this project took shape around an aspiration for a multicultural Los Angeles whose residents would have greater cultural understanding, achievable insofar as the group could bring younger and older people together through hip-hop and orchestra. The presenters rely on their assertion that the "audience will depend on what you program" to draw specific groups to performances and to bring different groups together. Hence through reified genres of hip-hop and European classical music, daKAH facilitates the presenters' project of encouraging intergenerational understanding at the public concert. Though it is seen how hip-hop and orchestra are "undone" in practice, they are reaffirmed for this project. In this context, the significance of daKAH's genre lay primarily in its social meanings, as particular musical and social aspects became the focus over others for bringing youth to California Plaza,

Figure 7. DaKAH at California Plaza. Photo by Guilherme Rafols.

supporting a project of intergenerational understanding and invoking a diverse Los Angeles.

For Grand Performances, a project of intergenerational understanding has a common logic to cross-cultural efforts. Age is added to the list of categories of diversity. "We know," a grant application maintained, "that we cannot only program for the usually older committed audience, but must find ways to build bridges to other cultures (sometimes defined by ethnicity and sometimes by such characteristics as age, socioeconomic group or lifestyle)." The success of these aspirations requires stereotyping taste by age. Young people, in seeing "their music" on a downtown stage, are intended to feel included in the civic body. The presenters also hope that older people, whom they consider their general audience, will have greater tolerance for young people by gaining an understanding of their music through daKAH. This intention, articulated by the presenters and Double G, is compatible with Grand Performances' general project of creating diverse audiences; however, rather than being marked by skin color, in this case audience members are marked by age. Validating Grand Performances' aspirations, on a survey given during daKAH's concert, a self-identifying eighty-one-year-old woman wrote that "public concerts are extremely important" because they "open minds." On hearing this, a member of Grand Performances' staff exclaimed, in a comment echoed later by Double G, "That shows we're successful! Where else would an eighty-one-year-old woman come to see daKAH?"

Tolerance comes in a double, interrelated package: a recognition of and respect for difference and a belief in universal humanism. The former, locating difference in the form of styles of music or dance, emphasizes increasing cross-cultural (or intergenerational) understanding on the part of the audience members who do not identify as listeners of the music presented onstage. The latter encourages a feeling of commonality with fellow audience members. Requiring only the performers and the music, for the first, seeing the strange, unfamiliar, or disliked art form presented on a professional stage is intended to give it respectability, encouraging the audience to value the art form in itself. The second depends on an audience comprising people identified as connected to the performer through shared background and others who are not; it is achieved by watching the performance with people of different backgrounds, including, and maybe especially, those who have a positive connection to or affiliation with the art form through taste or nationality. Through this experience it is hoped that people

might understand that they are all humans sharing the experience of enjoying the music. The aspiration to facilitate the second aspect by programming groups such as daKAH depends on bringing people who are different,
in this case younger, than the "older committed audience" and by showcas-
ing "their"—the "different" group's—art form on Grand Performances'
stage. In the case of hip-hop, it is assumed that the art form is disliked, or
at least unknown, to the older audience. Similar to world music, which
often has a pop element that makes it accessible to Euro-American ears,
daKAH packages the marginalized form of hip-hop with European classical
music, a genre that is understandable and accepted by those who, it is
hoped, will become more socially tolerant by listening to the music.

Age, however, has unique social characteristics and meanings that make
it function differently as a category of diversity from ethnicity, nationality,
or race. In the context of hip-hop, shifting to generation as the measure of
diversity makes it a substitute for race (Rose 1994; Schloss 2004). Preclud-
ing the presence of a stereotypical hip-hop audience that is feared for its
potential to instigate violence, presenters occlude race, demarcating it as
something that might provoke conflict. While older people and seniors are
considered "safe" in the context of low-income housing, youth are a source
of much anxiety for those same institutions. Yet youth is a historical category, and "the cultural meanings and social attributes ascribed to 'youth'
have varied a great deal across time and space;" youth as a site of both
promise and peril is an "offspring of modernity" (Comaroff and Comaroff
n.d.:1, 2). The Comaroffs argue that the exclusion and marginalization of
youth have increased as an aspect of neoliberalism; "as post-Fordist economics recast relations between capital and labor, profoundly altering
global geographies of production" (n.d.:6), youth emerge as a group of
people who can consume but not participate as producers or civic subjects
(n.d.:22).

Others find, in the marginalization of modern youth, the potential for
and signs of resistance, whether through style or the formation of counterpublics (Hebdige 1981; Negt and Kluge 1993). Some theorists of globalization see international youth movements as a hopeful sign that progressive
politics have not been completely eradicated by neoliberal capital (Sassen
1998). An emergent global hip-hop movement partially organized as the
"Hip Hop Nation" is one moment of this potential; hip-hop from the
United States, now almost fully commercialized, has provided indigenous
youth around the world, from Brazil to New Zealand to the South Pacific,

with a notion of "blackness" that provides a means of understanding their own marginalization, a form of solidarity, and a basis for claiming group rights (Hapeta 2003; T. Mitchell 2002). For Grand Performances, youth serve as the maligned category, understood as excluded from the "general public" and civic body and in need of inclusion in both as public concert audience members.

In articulating social aspirations of intergenerational understanding, particular musical aspects of daKAH are highlighted. Double G supports Grand Performances' project insofar as he aspires to bring hip-hop to European classical music audiences and bring classical music to hip-hop. He glosses somewhat stereotypical understandings of musical form with social groups, arguing that "we're not only exposing their culture to our culture, but exposing our culture to their culture. Hip-hop's a dance form, just like a suite, just like a minuet." He finds overlaps between the two genres in dance forms, though he uses examples from particular historic moments to define classical, omitting both early and contemporary European classical music in the process. For the Disney Hall performance later in the year, Double G was working on a composition in strict sonata form. This, he argued, would make classical listeners have to accept daKAH: "Whether or not you like it, you have to accept it, because we did it to the same form." On the other hand, he continued, "Right now, hip-hop could use a good lesson in form, I think. Especially the variation part. So by the same token, we give them the beat by giving them the form, and conversely, we give them the form by giving them the beat." Ultimately, Double G says, "What I would like to get is a sound that sounds really danceable and groovy, but at the same time the technique to generate it is really demanding of the musicians." In creating a space of cross-germination, Double G indicates how genres become defined through their legibility for and effects on the audience (hip-hop for classical audiences) and musicians (classical for hip-hop musicians). For the purpose of a project of intergenerational understanding, hip-hop and classical are defined narrowly, while jazz and other musical elements are put to the side, as they lack social meaning in this particular moment.

DaKAH's performance was compatible with other efforts on the part of Grand Performances to bring different groups together through music. A double billing of a Japanese shamisen player and African singer was intended to "mix up" the audience. The virtuoso shamisen player Agatsuma was paired with Afro-pop singer Ramatou Diakité, with the idea that the audience would see stringed instruments from two traditions, two parts of the world. The hope was that audiences who came to hear one with which they identify because of nationality or race would stay to hear the other. The success of such events is evaluated through visual recognition of audience demographics and mailing list sign-ups. Most of these kinds of events—which also include festivals of performers from different countries within geographic regions—usually fail only insofar as not as many audience members stay for the whole evening as would be hoped. That daKAH's performance was a hit with board members and at least one eighty-one-year-old made it a success in these terms. However, a similar effort—the pairing of the Nortec Collective and the Kronos Quartet—presented a near institutional crisis, ultimately reflecting the particularity of the public as the tastes of one group supported the exclusion of another. Just as the ostensible universality of the democratic public is created through music, so are its inevitable limits.

The evening was part of the Latin Alternative Music Conference, an annual showcase of "rock en español" and related music, panel discussions, and industry hobnobbing. Kronos Quartet's *Nuevo* CD, featuring work by Mexican composers, had not sold as well as had been hoped for. This concert was considered an opportunity to expose a Latin youth audience to the Kronos Quartet and provide the Kronos Quartet with a new, CD-buying audience. Plankton Man, a member of the Nortec Collective, had done a remix of one of Kronos's pieces that was included at the end of the CD. The members of Kronos were excited to play this concert. They are generally interested in new music and popular music from around the world but do not often have the opportunity for such a billing. Programming the Kronos Quartet was compelling for Grand Performances, which does not often present groups that are so well-known. Because of Kronos's status as a contemporary classical string quartet, the concert became a fund-raising opportunity. A bank with offices in a nearby building, represented by a board member who is also a cellist, took this opportunity. The bank's logo went in the program, it was thanked at the beginning of the concert, a banner was hung with its name on it, and seats were reserved for its employees.

Because Kronos was more famous it was featured during the second half of the program. Nortec Collective started the concert shortly after 8 P.M. Onstage, four men stood behind laptop computers placed on a table. It was still light out and the music was not as loud as it would have been in a club. The technical director intentionally kept the volume lower so the performance would be compatible with that of the Kronos Quartet. With funds tight for the season, Grand Performances had declined to pay for the projection screen that normally provided the Nortec Collective's perform- ances a visual component. Young people milled about, settling down on the granite plaza floor in areas without chairs. Soon the complaints began. Some of the older, regular audience members did not like the music. Donors, tired after work and not hearing the music they had come for, left. The director, who fielded these complaints, was upset that the group had been misrepresented in the brochure. The text describing them made it sound as if they would be playing familiar instruments like the accordion, tuba, and drums, but here they were, not "doing" anything that could be seen. Someone decided to turn the fountain back on, which shot water into the air behind the performers, so at least there would be something to see. Value judgments were made. An older man who reported that he was between forty-six and fifty-five and white came to the information booth and wrote on a mailing list sign-up form, "Nortec Collective are boring. They're not creative. Why are they here?" Another ("female, over 56, Cau- casian") wrote "1st act 8/15 monotonous." For those complaining, it was not personal taste that was at stake but rather a fundamental evaluation of good and bad music.

On the other hand, young people came up to the information booth and exclaimed about how great the place was. If they had come to hear the Nortec Collective, they stayed for Kronos. Excited by the venue and performance, many signed up for Grand Performances' mailing list. They recorded their excitement in comments such as, "Never knew about this in my 14 years of living in L.A. It's great." "This is very cool!" "¡¡¡Buenis- imo!!!" "This is great! I love tonight's performance 8-15-03!" And "8/15—good combo." Most of these comments came from people who were between twenty-six and thirty-five years old and who were attending a Grand Performances concert for the first time. For this new audience of younger people, the concert was a success. The Kronos Quartet was happy that, unlike at previous performances of this program, the audience knew the composers and applauded when their names were announced. But for the other, older audience who did not like Nortec Collective, the concert was a huge failure. And at this concert it was the latter who affected the response of the presenters. An air of crisis pervaded the office during the week following the performance. Meetings were held behind closed doors. Staff whispered that this kind of program would never happen again.

As with daKAH, in pairing the Nortec Collective and the Kronos Quartet, the presenters emphasized diversity of age rather than ethnicity or race. In emphasizing age, Grand Performances hierarchizes the respective genres and taste publics. Though it is hoped that coming together in harmony will foster a sense of agreement and consensus that will extend to the city at large, the two groups are not on even terrain. Diversity of age frames the older group as Grand Performances' regular, committed audience. This group holds the keys to legitimacy, which, when supported by Grand Performances' staff, reflects the limit of the purported general audience of public concerts. Hence more broadly, correlating music with bodies marked by ethnicity, race, or generation marks a hierarchization of these groups that is also a means of making a general public into which the excluded or maligned are incorporated. The limits of the general public expose the construction of this structure and the necessity of certain shared identifications at the expense of others. A person's presence facilitates harmony and understanding only if she fits the requirements of both recognition and participation. Cases that threaten institutional stability indicate how the consensus of the general public is reasserted as power is wielded in closed-door meetings that reestablish the framework through which aspirations for civic diversity might be realized. Limits, ruptures, and uncompleted gaps reflecting boundaries of the public are re-closed through exclusions of those subjects and musics that disrupt the "general" consensus, now clearly partial and particular.

"An L.A. Phenomenon"

Insofar as diversity serves as a defining—and exceptional—feature of Los Angeles, daKAH embodies the city through musical and social diversity. DaKAH's diversity resides with the performers, who do the work of invoking the city as mixture, as crossover, as geographic, and as hip-hop. An MC draws on the Gang Starr song title "Code of the Streets" to create an original verse that situates daKAH as in and of Los Angeles. He articulates tropes of mixture and diversity, making a lyrical metaphor and literalizing a musical and social correlation that maps daKAH into the city of Los Angeles. The orchestra plays for eight bars, with the MC interjecting, "Uh." At the ninth bar he raps:

Well it's the code of the streets
The code of the ghetto
I used to say hello to my fellows
That's stuck in Meadows.
I'm from One-Twenty-Seven Street and Normandie
If y'all know about that then y'all are feelin' me.
Los Angeles the land of the scandalous
Get your hardness banished with this.
DaKAH, can you feel this, sixty pieces
We all up in you like streets is.
On the block a Thomas Guide
'Long with that we provide.

By using Gang Starr's song title in the flow of the lyrics the MC acknowledges his debt to the original composition and signals to the audience what
song is being performed. However, rather than a glamorous description of
the riches gained from gangster life as provided by Gang Starr, he emphasizes the geography of Los Angeles, referencing the *Thomas Guide,* the thick
spiral bound map L.A. drivers carry to find their way around what is often
an unfamiliar part of town. He connects the geography of the city to the
members of daKAH, who, with different musical and social backgrounds,
have come from around the city to play together in downtown Los Angeles.

The lyrics place daKAH in Los Angeles, situating the MC in a tradition
of Los Angeles hip-hop which, as Shaw writes, "Since its beginning . . . has
reveled in listing specific locations, street intersections, schools, stores,
parks, housing projects, gang turfs, and hoods" (2002:14).[3] Identifying a
street in South Los Angeles connects the rapper to the sociospatial history
of hip-hop in Los Angeles. Gangsta rap, which put Los Angeles on the hip-
hop map, emerged from the post-industrial terrain of South L.A., a site,
like so many other African American neighborhoods around the country,
of racial segregation and a vacated economic base (Davis 1992; Kelley 1994;
Neal 2004; Rose 1994; Shaw 2002). Gangsta rap's lyrics, renowned as violent
and misogynistic, are described as in part reflecting the reality of the neighborhood while at the same time exaggerating such claims for reasons of
discursive competition (Kelley 1994). This music and its lyrics were part of
an Angeleno city-making process, describing a neighborhood for millions
of fans worldwide and affecting how local law enforcement officials
responded to the population.

The multiscalar geographies presented in this verse connect the personal with the urban. As the MC who authored these lyrics explained, the intersection of One-Twenty-Seven Street and Normandie is where he grew up and where two generations of his family lived before him. The intersection is also in the middle of an area of the city where the infamous Los Angeles gang the Crips was formed. He moves from the specific, delineated geographic area of that intersection to invoke the city as a whole, "the land of the scandalous." The orchestra, with more than sixty musicians, comes from around Los Angeles. Each musician, like the MC himself, is from a specific street, with its own significance on the scale of the street or the intersection. Thus the audience members (the "you" of "can you feel this"), in hearing daKAH and feeling the music in their bodies, are sonically imbued with the geography of the city in its components of streets. A multiscalar geography from the audience member's body to the space of the performance to the streets of L.A. and the city as a whole is produced through sound ("we all up *in* you"), street names ("One-Twenty-Seven St. and Normandie"), and the tome of Los Angeles's spatial representation, the *Thomas Guide*. The deployment of these spatial scales emphasizes the present moment of the performance, stressing, in the end, the relationship between the performers and the audience, as the personal experiences of the musicians in their neighborhoods intertwine with audience members at the concert to produce the city.

With daKAH, the multiple ways in which identifications of diversity are assumed, invoked, and made through musical composition and performance contribute to a continual process of connecting sound, space, and the social in a frame of diversity. As Stokes advises, "Musical performance is multitextual, embracing all manner of contradiction. . . . Its meanings can never be fixed conclusively in spoken language, and the 'same' music can . . . be appropriated by different groups for quite different reasons" (2000:216). For diversity to stand as the dominant frame of interpretation requires closing gaps opened by the potential of multiple meanings. MCs express the musical meanings in play at Grand Performances' events as their lyrics suggestively articulate the process of closing intertextual gaps (Bauman 2004; Briggs and Bauman 1992). Ultimately, they suggest how in bringing people together, civic performance fosters spatial proximity through music, evoking broader social aspirations of multicultural community.

Meganut: Dynamics
    The dynamics
Hey Gabby?
Gabby: What's up, Meganut?
Meganut: Yeah, this daKAH y'all
    This daKAH
    Make some noise for daKAH Hip Hop Orchestra
Gabby: Right, right
    It takes more than one or two or three or four
    It takes a lot more people
Meganut: Everybody
Gabby: Come on
    It's the dynamic
Meganut: The dynamics
Together: The dynamics
    The dynamics of a sound body and soul

The lyrics move from music to people to music in describing the energy created by the relation between them. "The dynamic" and "the dynamics" invoke both social relationships and sonic volume. The assertions are incomplete: "it takes a lot more people" to what? The audience must draw the connections and fill in the gaps, which they do by recognizing an ideal L.A. at public concerts. They continue,

Out of control,
In the flow frequency,
Frequently vibrating around the cycle that brings us together,
Magnetized by each other's imagination,
A station picked up by a broadband bandwidth,
Band aid capabilities,
"I am" [from the DJ/sampler]
Capable of healing with feeling

Shifting from sound to body and back to sound, the lines stress the relations between the two by interweaving the references: "The dynamics / 'Til the sound" moves to "Body and soul / Out of control . . . Vibrating" and back to "bandwidth," as movement ("vibration") creates collectivity ("bandwidth"). The vibrations of the people and the music suggest an experience of losing a sense of control while dancing, letting oneself be in a "flow" and evoking the "higher level of relatedness" described by an audience member. The "more than one or two or three or four" people become unified and energized by each other, "vibrating around the cycle" and "magnetized by each other's imagination." As social collectivity is produced through music, "a station picked up by a broadband" moves the experience into a mediated and potentially discursive realm that might articulate the meaning of this experience for others or facilitate an understanding of this experience in terms of broader social concerns.

> Come and feel these future beings,
> Being the future,
> Seeing the past pass us by,
> Open to all possibilities,
> Unseen, unheard and woven in the fabric of our fabricated thoughts,
> Do y'all give a nut,
> Give a what,
> Give a nut!
> Give it up y'all
> daKAH!

The MCs call on the audience to move into the future along with the beat, which marks the linear unfolding of music in time. Repetition, referring to its own near and distant past, allows the future to be predicted by the regularity of the beat; yet only in being with the beat as it continues can one know whether that prediction holds or not. As the audience members stay with the beat they feel and are future "beings." The past passes, carrying with it things that could have but never happened, unimagined in thought. The future in which they are present is embodied, while the past is outside of them and moving by; unseen and unheard, its possibilities remain unknown. Openness to the music is an openness to that which is outside one's self but that constitutes social relations. As an aspiration of public concerts, the future is people coming together; as a utopic aspiration, its potential is realized in experience. In finding these ideals in the future, however, they become unlinked from their particular histories, appearing abstract and universal, like music, and like the higher order of multiculturalism—that of sameness in difference—aspired to at public concerts. Music resonates within even as it is separate from the body. The past, passing us by, suggests something lost or let go, while the future becomes a space of creativity and unknown possibilities. Gabby brings the general description of music and people back to the present of the performance and into the group by referencing her partner's name ("nut" for Meganut) as she directly addresses the audience, exclaiming, "daKAH!"

A male MC continues alone, engaging the audience in order to realize aspects of the verse. He gets the audience to yell, saying, "Put your hands together. . . . Now I need some assistance." Connecting music and body as both personal and open to interpretation by the audience, he raps, "My movement is my music and I'm usin' it however you choose it." In participating in public concerts, performers, audience members, and presenters work to close the gaps of incomplete verbal references that evoke ideals of community and diversity, of identifications linking audience members with musical style, and of the connections between embodied experiences and civic ideals. These gaps are not always closed successfully; moments of rupture or failure and resulting institutional responses suggest how the order of a general public is produced and maintained. Reflecting the dystopic side of this utopia, consensus and agreement require the exclusion of dissent, and of bodies and sounds that do not meet the requirements of those defining the terms for a multicultural sociability. Nevertheless, in the end, gaps are closed sufficiently, and a diverse city is perceived in performance.

The MC who begins Gang Starr's "Code of the Streets" writes musical diversity into the geography of Los Angeles; his lyrics describe how the orchestra members, in coming from around Los Angeles, personify the city, sonically entering the bodies of the audience members and thus sounding the city at California Plaza. An online interview with Double G places daKAH in its home city, starting off with the statement that "Dakah is an [sic] Los Angeles phenomenon" (Oshun n.d.). Echoing the sentiment of the reviewer, an audience member said that the evening had felt "very L.A." because of the "slightly cheesy atmosphere" of the Plaza and music that was a mixture, combining strings with hip-hop. In a practical sense, daKAH is very much in and of Los Angeles; organized in the city by musicians who had come to a place where they could both find work and play music they love (which is sometimes the same thing but often is not), the orchestra is now too large to tour without having a large amount of funding. Moreover, daKAH provides a musical example of diversity that evokes a parallel with a diverse Los Angeles. Emphasizing the work done by the performers as the basis and focus of musical meaning, the diversity of the group and the musical styles makes daKAH "uniquely Angeleno."

Conclusion

Diversity as something "uniquely Angeleno" frames civic performances of groups like daKAH in the context of the city. In turn, this framing crafts a notion of the city as defined primarily by its diversity. DaKAH's diversity, organized around multiculturalism, urban geography, and music, is writ in the same terms as the diversity of the city. Hip-hop orchestra and Los Angeles converge, sounding and re-sounding one another as they are reflected back to each other in performance. The emphasis on diversity reflects a neoliberal thrust to eliminate class politics and other "divisive" issues from public discussion. Yet even as describing daKAH as reflecting Los Angeles through a trope of diversity shifts from the geographic, racial, economic, and musical history of Los Angeles hip-hop to an evocative field of association, hip-hop's social basis is not lost. As daKAH shows, diversity also includes social and economic inequalities, urban violence, popular culture, tensions of multiculturalism, and emergent racial politics, all of which are emblematic of the global city.

Hip-hop as a music of urban America and as a global export has now taken on a life of its own as a mode of youth expression from Cuba to Ghana, Australia to Japan (Condry 2006; T. Mitchell 2002; Shipley 2007). For Angeleno musicians, participating in the global popular culture industry is one aspect of the ways in which processes of globalization are situated, produced in sites that are at once local and global. Thus daKAH, as a group that is in and of Los Angeles, is embedded in the ways in which global city formation is multiscalar in practice, as inequality, racialized urban geographies, and diversity—all shaped in part by neoliberal capital—intersect with a global popular culture form that is very much of Los Angeles.

4

Sonic Civilities

Music is . . . a way of perceiving the world.

—Jacques Attali

Thank you for playing my music. I'm from Veracruz.

—Grand Performances audience member

El Vez begins the second half of the performance dressed as Uncle Sam, with white vest, huge white bell-bottom pants, and billowing satiny blue sleeves. His band, the Memphis Mariachis, wears straw hats and white shirts. An American flag hangs at the back of the stage. "El Vez for Prez" opens with a rendition of "God Bless America." Between songs, screens on the side of the stage show videos of people—purportedly audience members—asking El Vez about his position on various issues. Taped beforehand, they are framed as a performance of direct participatory democracy. "What is your position on immigration?" one asks. "I love it!" El Vez replies, "The country is built on it! Viva la raza!" "What do you think of the problems of the inner city?" "El barrio?" he asks, launching into song.

> As she shoos the flies
> On a hot and smoggy summer morn
> Another brown little baby is born
> En El Barrio [The Elvettes echo: "En El Barrio"]
> La Mamacita cries
> Porque there's one thing that she don't need
> And that's another hungry little vato to feed
> En El Barrio

"El Barrio," set to the tunes of—as a rockabilly fan explained—"In the Ghetto," Traffic's "Dear Mr. Fantasy," and the Beatles' "I've Got a Feeling" (Washburn 1994) continues,

> So he joins a gang
> Porque there's one thing that he can't stand
> And that's to have to join a mariachi band
> En El Barrio

Until the boy steals a car, and the mother, in despair, decides to move

> Out of east L.A.
> With no more gangs and no more crime
> To the promised land out in Anaheim (near Disneyland)
> En El Barrio

El Vez rails against President Bush and the Patriot Act. When his backup singers, the Lovely Elvettes, rip posters of Bush's face in half the audience screams in approval. True Americana, the evening ends with fireworks.

Representation, recognition, and participation, tenets of democracy, structure public concerts as civic performances. At civic performances these terms are at once performative and political, organizing cultural identity and civic membership. El Vez's performance sounds civic space by performing the intersection of cultural identity and electoral politics. Touching on subjects ranging from immigration to the Patriot Act and life in East L.A., he raises issues pertaining to overlapping civic, national, and international politics with humor and spectacle. Sustained throughout is an underlying expression of Chicano identity politics that expresses how, though the American condition is built on inequality and exclusion, there is hope for change and greater inclusion (Habell-Pallán 1999; Rubin 2004). Identifications are drawn along multiple axes as cultural identification and political identification converge, made accessible and inclusive to all through the familiar tunes of Elvis songs and other popular music.

Five years before, at El Vez's last performance at California Plaza, some audience members came dressed in rockabilly attire, presenting a subcultural style of Latin rockabilly; other men sported large sideburns and one dressed as Elvis. That year's theme was a quinceñera party, and fifteen-year-old girls in white dresses and tiaras ate cake onstage, performing a Mexican

American tradition to celebrate El Vez's fifteenth year as the Chicano Elvis. I worked in the information booth during the performance, and streams of people came by to sign up to be on the mailing list. In the space asking them to "please describe your ethnicity," many listed Chicano/a, though some wrote Mexican, Latin, or Mexican American. A few self-identified as Asian. Whites responded with "white boy," "blanco," and "it's rude to ask." "Viva La Raza" was sung to the tune of "Viva Las Vegas" while El Vez waved an American flag. The call, with its accompanying fist raised in the air, was raised without disruption. Though at other events it was considered divisive when screamed from the audience, here it was laughed at. Humor, spectacle, and Elvis allow things to be expressed that might otherwise be curtailed.

The civic of public concerts is a space of consent, a space safe for difference and a space where difference is made safe. Music organizes and supports the terms by which representation, recognition, and participation transpire, serving as the medium for the signification of identities. Music framed as universal follows the structure of liberalism, the political foundation of the United States, which maintains a "tension between particularistic 'I's' and a universal 'we'" (Brown 1996:153). Personal identification with particular musical forms creates commonalities under a sign of universal "music." "My music" extends identification to others, such that at the public concert the performance of identities becomes a means of connecting diasporic groups with an imagined multicultural Los Angeles in which diversity serves as a form of belonging in the city. Made specific to Los Angeles, civic ideals of public concerts emphasize the creation of a multicultural audience that reflects the diversity of the city. Learning about another culture through its music at a public concert projects empathy and understanding toward audience members whose music is being performed as well as others who share a cultural attachment to the music. Sound "figures in bodily ways of knowing and being in the world" (Feld 2000:173), framing and providing a foundation for emergent senses of belonging. And in coming together in the downtown space of the corporate plaza, the segregated city might be shaped according to presenters' aspirations, becoming unified and harmonious.

Public concerts are sites of "identity-in-process," where identifications around musical taste and practices help produce a multicultural, global city. Translocal civilities are produced through identifications of musical meaning that attach performance to nation. Though the state is significant in its ability to fix categories, their making is part of wider social and political processes. Music creates a space for identifications not allowed by the liberal state, which, in positing universal man through the language of equality, creates difference through its exclusion from the sphere of the political. As Marx argues in "On the Jewish Question," these identifications, the "substantive conditions of our lives" (Brown 1996:153), which, not allowed to enter into an abstraction of the universal "we" of liberalism, are allowed "to *act* after *their* own fashion . . . and to manifest their *particular* nature" outside the state (Marx 1978:33).[1] Belonging organized around multiculturalism—predicated on a notion of difference—can be negotiated in spaces such as civic performances. By avoiding the polemics and politics that emerge around the tension created when rights claims on the basis of difference are made on a state whose logic is universal man—the most visible form of which in the United States is affirmative action—civic performances create a space for the production of multiculturalism and of a diverse city. The terms of representation, recognition, and participation as manifest in this context do not necessarily have bearing on the political strictly defined. Yet recognition of the categories of multiculturalism reflects the ability of civic performances to shape the terms of liberal democracy outside the political proper even as it marks their imbrication in wider regimes of governmentality (Connolly, Leach, and Walsh 2007; Fraser and Honneth 2003; Hegel 1991; Kymlicka 1995; Markell 2003; Prendergast 2000; Taylor 1994). Recognition of difference, a space for multiculturalism, and identification of cultural expression as ways of understanding social groups are caught up in a nexus of performance and the politics of multiculturalism.

At civic performances there is constant negotiation over the relationship between performance and politics. As a space of privatization, a public-private partnership, Grand Performances' public concerts continually shift registers, using the language of democracy to support their project in the name of the public. National politics glossed as culture are rejected as constraining by artists who choose between being "bound by recognition" and the "freedom" offered by international art forms. Identity categories are configured by, though in excess of, census categories. Excluded from the abstraction of the political, civic performance reinhabits a space of abstraction by excluding politics on the basis of their divisiveness. The consensus wrought from the exclusion of dissent also creates a political space in which there is consensus about recognizing identity in performance, about the possibility for performance, whether music or dance, to signify identity. The particularism of the "we"—the general public of civic performance and the city—is revealed in moments of contestation over identification or inclusion.

"My Music"

In feeling moved by "my music," one makes a sentimental connection to a place or event, which will resonate with poignant, personal meaning. With citizenship transformed by dynamics of globalization (Castles and Davidson 2000; Oboler 2006), emergent forms of belonging cohere in civic spaces that can encourage "interaction . . . across diversity" (Isin and Nielsen 2008:8). Attendance at Grand Performances' events shapes civic participation as a way of being in the world, as cosmopolitan identifications and forms of belonging are evoked by the public concert. The ability of music to serve as a medium for global, identity-driven, and affective forms of belonging is credited alternately to its sensory immediacy and to its prevalence in social life, to its particularity and to its generality, allowing it to situate a person in terms of discrete identities and transcend social difference.[2] As Frith suggests, "Music constructs our sense of identity through the direct experiences it offers of the body, time and sociability, experiences which enable us to place ourselves in imaginative cultural narratives" (1996:124). "My music" provides a site for identifications between self, music, and wider identity categories that might be national, affective, or sentimental.

The man who wrote on his mailing list form, "Thank you for playing my music. I'm from Veracruz," expressed a personal identification that invoked and reinscribed links between self, territory, and music. For others "my music" might be something that distinguishes the person from an assumed mapping of category and self, instead emphasizing a personal, affective connection with the music. One Grand Performances audience member laughed when I asked him what "his music" was. "Music that I listened to when I was young and I liked," he replied. "I don't know if there's any explaining why you like a particular artistic expression; it just resonates with you." Despite his disavowal, he acknowledged the presumed relation between music and identity even as his personal taste was in marked difference from what he described as his background. He continued, saying, "I'm an African American male born in the South, raised in the Midwest and South mostly, and no one that I know, you know, my family, likes the kind of music I like. I would listen clandestinely to the Beach Boys on my car radio in high school." Not being able to explain why one likes something that "just resonates with you" entails an unlinking of music and language in which the nondiscursive is also "beyond language," understood as more "real"; not able to be explained with words, it is marked as immediate and closer to the self and to experience (Hoeckner 2002). Echoing this sentiment, another audience member wrote on a survey, "I like contemporary jazz, R&B, funk, contemporary gospel, and world music" because "that's what moves me." In "resonating" and "being moved," the body becomes the site for that which is nondiscursive, the affective, subjective, personal, authentic, and true, and the location for the recognition of many of the collective ideals of Grand Performances.

"My music," whether national or sentimental, assumes a shared understanding of a relation between music and identity. A history of conceptualizations of culture organized around correlations of group, expression, and place that solidified with Boas's development of Herder's notion of *Volk* provides a foundation for ongoing understandings of the potential for music to signify categories of ethnicity, race, and nation (Born and Hesmondhalgh 2000; Radano and Bohlman 2000). Yet the relationship between music and identity, while conditioned by these philosophical and social scientific formulations, remains a process that is continually created and negotiated. A fixity of nation, ethnicity, or race that serves as the basis for the possibility of shifting identifications is produced at Grand Performances' events through, among other things, personal identification on mailing list forms, language use, and programming that assumes correlations between music and identity.

Identifications with music are conveyed by performance, affective relationships with the performance based on nationality or experience, and recognition of others' relationships to the performance that facilitate cross-cultural understanding. These identifications are articulated most directly through dress, modes of participation, and discourse expressed in informal and formal settings. "Music has a formative role in the construction, negotiation, and transformation of sociocultural identities" (Born and Hesmondhalgh 2000:31), and always already constituted "cultural" and "national" subjects are interpellated by their own identification with a particular musical tradition, by the identification of the signification of performance, and by their recognition by others. "Situational and contextual" (Brubaker and

Cooper 2000:14), identifications are made about one's self and others; "my music" can be projected onto others, sustaining correlations of music and identity that allow music to facilitate cross-cultural understanding and multicultural belonging,

Performing their identification with the music, audience members come to California Plaza wearing forms of traditional dress or T-shirts bearing the name of the country from which that evening's performers come, carrying flags of their shared home country, and speaking the language of that nation. Identity—as nation, ethnicity, or race—is embodied by the individual, whether performer or audience member. While generally based in personal experience and background, those who sustain a cosmopolitan sensibility through travel, language study, or media exposure share these expressions. A white woman on the downtown DASH bus said she often attended performances of African music with her partner, saying simply, "We like African music because we've been to Africa." A downtown resident and frequent attendee of Grand Performances' concerts had developed an interest in Indian popular music through Bollywood films, which she had sought out after watching *Moulin Rouge!* She loved the Indian classical music she had heard at California Plaza and was looking for recordings of similar music, saying, "I know it when I hear it. I know I like it." At a recent performance a couple in the audience that she described as "regular old white people from Orange County" who she thought had been to India had told her about an Indian newspaper with listings of area events. Based in identifications that attach to something other than primordial identity, cosmopolitan dispositions allow for shifting affiliations with relatively fixed national cultures.

When presenters address the audience in Spanish or Chinese, language interpellates an audience presumed to share a national linguistic community with the performers, assumed to be present for that reason. Audience members in turn voice their civic membership through language. Spanish—uniquely among non-English languages—is spoken in the public sphere of Grand Performances. While native speakers of other languages address strangers in English, Spanish is used readily to ask questions of staff members in the information booth, whether or not the speaker or addressee is a native Spanish speaker, with audience members continuing in Spanish even when responded to in English. Spanish-speaking performers often speak Spanish exclusively to the audience. With persons of Hispanic or Latino origin comprising nearly 50 percent of the county's population, this

Figure 8. Signs ready to be placed in the plaza before a Beijing Modern Dance
Company performance. Photo by author.

seems appropriate, though it is not entirely commonplace in Los Angeles's
civic spaces. Los Angeles's location in the Mexican-American border region
and its ties with Latin America through migration and trade dating to its
origin as a Mexican city now remade in a neoliberal service economy provide a basis for a cosmopolitan stance. When an African American man
addressed me, while I worked in the information booth, in American-
accented Spanish before walking off with his friends speaking English, his
effort appeared more as a means of marking his inclusion in an international, cosmopolitan space that might include listening to music, drinking
Pacifico beer, and speaking Spanish. At the same time, this kind of inclusion is limited, and a native English speaker, used to his position of centrality and power, said, "I never felt so marginalized as a non-Spanish speaker
as I do here."

Correlations between music and identity continue to be negotiated and
undone, underscoring their processual and contingent nature. At an event
about hapa identity (used to refer to people who are part white, part Asian),
an audience member was irritated when identified as hapa, as this identity
did not resonate with her experience growing up on an island where this
background was the norm. "I'm just like everybody else," she wrote when

she grudgingly agreed to be part of the photo album of that evening's audience. Happy to have a family that came from different parts of the world, she felt it was no different from having part of your family be, in her words, "butchers." The most memorable performance for her was by a contemporary hula group, not because it was Hawaiian but because the dancers wore red, her favorite color, and danced to "Ever I Saw Your Face." This refusal of identification pushes at the totalizing thrust of correlating performance and identity. However, contests of identification, limits to consensus, and resistance to being identified by others put pressure on, but do not undermine, the possibility for a diverse Los Angeles to be identified at Grand Performances' events, which remains as the most consistent narrative and framework. Enabling the possibility for a general recognition of a diverse city to emerge at public concerts, "identification does not *require* a specifiable 'identifier': it can be pervasive and influential without being accomplished by discrete, specified persons or institutions. Identification can be carried more or less anonymously by discourses or public narratives" (Brubaker and Cooper 2000:16). At public concerts, the power to identify and define is dispersed, and for the most part presenters, performers, and audience members express shared understandings of musical meaning and multiculturalism.

An understanding of music as universal allows personal identifications of "my music" to be extended to others. A woman who had identified herself as African American on a survey while attending a performance of jazz vocalist Jimmy Scott described how seeing black people onstage filled her with joy. She elaborated, stressing the significance of the diverse audience with whom she had been sitting: "I think that's true for all ethnic people, to be able to hear the music of their culture and look around and see an audience of all variety. I think it helps people come together, to see that there's no color on beauty; it crosses color, it crosses features." The idea of music as a motor for bringing people together and overcoming differences took shape in the Romantic era of nineteenth-century Europe, perduring to the extent of becoming cliché. The European Romantic period saw a new consideration of musical harmony as a reflection of social processes. Musical harmony "turned listeners inward to produce absorbed silence and attention. Socially, it made the hall a metaphor for the street" (Johnson 1995:280). An American tradition of public concerts during the Progressive Era sustained these musical and social ideals (C. Smith 2007; Vaillant 2003). In 1919 a commentator reflecting on music as one of the

"Ideals of America" wrote that "Music has one power which no other form of art expression possesses in equal measure, that of bringing a mass of people under the spell of a single uplifting emotion and thereby welding them together as a unit in the fellowship of a common experience" (Dickinson 1919:228). Assuming ideals of American social life through musical meaning, this position stresses the ways in which music supports a common, affective experience that enables listeners to find unity in diversity.

Grand Performances maintains a notion of music as a universal form that has different sounds, styles, or techniques in different places and as practiced by different people. Framed as a "universal," music supports the articulation and transcendence of difference (Kapchan 2008). Grand Performances invokes the universality of music when deploying current musical forms and social dynamics to impart broad social aspirations. A 2002 program for the screening of *Scratch*, a documentary on DJs, read, "What is most evident in *Scratch*'s message is the universal language that flows from this art form." The enthusiasm of the featured musicians in participating in the documentary could be seen as evidence of their willingness and interest in sharing their music with viewers. The presenters ascribe a notion of universality as part of their project of making connections between viewers who might be skeptical of the merits of DJing and turntablism as an art form or frown upon the musicians portrayed. Ventriloquizing the musicians whose practices come via the mediation of a film, the program notes assert, "The musicians clearly understand this and have no qualms about sharing their love of their music with anyone who has an open mind to hear it. We hope," the program concludes, "these sentiments are echoed by our faithful audience." Expressing clearly their own understanding of music, the presenters use the musicians as a moral example for the audience. The musicians, they assert, are open-minded and generous, willing to share their universal gift of music with anyone. The audience should meet this generosity of spirit and acknowledge the higher order of a "universal language," feeling, at the very least, tolerance for the music and its performers, insofar as it might be different than what they are accustomed to.

An audience member who identified as an older, middle-aged Japanese American man said he listens to "'world music' such as Japanese, Indian, Indonesian because they have for me a universal appeal, elegance, sophistication, and can be related to my own personal life by hearing the experience of others through songs." Moreover, he felt that cross-cultural understanding achieved through Grand Performances' programming would extend to social interactions in the city: "With our diverse, Balkanized population, we need venues where we can sample, peruse, experience (no matter how superficially) other cultures, opinions, ideas, customs, etc. More contact means we desensitize ourselves to our differences and hopefully see more similarities. We need to encourage more ethnic/cultural tolerance." Hence a disposition of openness and tolerance brings these processes of registering and overcoming difference into the city, structuring ways of being together in the city. At a performance of Indian classical dance a man from Oaxaca said simply, "It's important to learn about other cultures through music."

Performing Culture

Identifications with music, whether of one's self or of others, are collective, interactive processes, imparted through dress, disposition, epidermal signification, and verbal communication. Post-performance question-and-answer sessions called Music, Movement and Meaning (MMMs) facilitate the verbal expression of audience identification and interpretation. Drawn from experiences of listening to and watching a performance, interpretation invokes a range of social categories that help create musical meaning and signification. Talk about music is always an act of interpretation, a "process of intuiting a relationship between structures, settings, and kinds of potentially relevant or interpretable messages" (Feld 1994:85). Presuming relationships between music and nation as the shared frame for interpretation, audience members muster knowledge of national culture and politics in their interpretations of a performance, presenters program performances whose content is intentionally legible as such, and performers straddle a line between these demands and a commitment to abstract form. Not requiring agreement in interpretation, it is sufficient that there is consensus that meaning is legible through performance. And while social identifications around music do not exhaust the meanings made at Grand Performances' events, they reflect a general tendency heard in talk about music and dance and seen in modes of participation, ultimately supporting the recognition of the diverse and global city at the performances.

MMMs, a season brochure explains, "give everyone participating the opportunity to share reactions to the presentation and to discuss the different 'messages' that people get from these performances. In a community as

Figure 9. Audience members at a performance of Ilgi, a Latvian folk ensemble. Photo by author.

diverse as Los Angeles, it is inevitable that different people will 'see' different things in the performance."

> Each of us carries a different set of experiences, different family stories and different cultural histories. Each of us discovers different "truths" when we view provocative performances. . . . Different perceptions of truth are a part of our community—a community without a single majority population. Hearing the different "truths" may help us better understand each other and may help us work together to make Los Angeles an even better place to live and work. (2001 season brochure)

MMMs emphasize contingent truths drawn from experiences of identity and difference that nevertheless ultimately promote an ideal of universal humanism. Identity, especially if different, is considered something that should lead to harmony between people. It is not a source of political struggle, disagreement, or violence. This harmony, learned and experienced at the public concert, can extend to the city as a whole, making Los Angeles a peaceful place because its residents understand each other's different "truths." Thus, in the end, the "general human being" emerges as the key to harmony, who, nevertheless, is writ as particular, with a singular perception coming from a unique background.

When Grand Performances' director began MMMs with a story about how he had interpreted that night's performance based on his skin color or gender and encouraged audience members to ask questions related to "where they were coming from," there would be an awkward pause before people began asking questions, which then seemed to consciously avoid the topic of identity. In 2003 the director stopped introducing the MMMs this way. The grant intended for MMMs had not been renewed, and there was no longer pressure to fulfill these particular aims. Since then, however, audience members frequently raise questions that focus on issues of identity, asking the artist how his or her background influences the work and how national concerns come into play in the performance. MMMs are a privileged space for the expression of interpretations about form, social context, and meaning. Though meaning, particularly in terms of audience identifications of performance content articulating something about national culture and politics, is not necessarily a primary point of inquiry, MMMs are the space in which it is discussed most often. MMMs are held after events that the presenters consider thought provoking; often these focus on issues of identity rather than form, though the latter also happen. During this time certain discourses are encouraged over others, maintaining consent by excluding statements that might challenge the agreed-upon terms and content of discussion. Hence, when after a particularly charged performance addressing international responses to clitoridectomy a man argued to include male circumcision in the discussion, the moderator closed off discussion by responding, "It's different."

Music stands for the meaning-making medium, and interpretations and identifications made possible by its framing can be extended to other mediums such as dance and spoken word. With music also having residue of its status as the "'pure' art par excellence" that "says nothing and has *nothing to say*" (Bourdieu 1984:19), presenters consider these other mediums to provide clearer expressions of the possible sets of meanings. According to Grand Performances' director, modern dance is programmed for its ability to convey stories from other countries in ways that are more legible to a general audience than are other art forms. Positioned by presenters and performers as a neutral, abstract form through which particular content can be expressed, modern dance draws on the logic of the universality of music. Audience members rise to the occasion, asking pointed questions about political content in dance performances. Drawing on their understanding of the country from which the dancers come, they asked the director of the Guangdong Modern Dance Company, "I saw militaristic, nationalist overtones. What do you think about the global situation?" Another asked, more openly, "Do you have a message? Does it mean something? Is there a message behind the dance?" The company director responded to these questions by saying, "The first piece was a comment on society at the present. Dance is a free form for the artists to express themselves. Today, the works are about how we see our tradition. Some works are more political. Some are about Chinese daily life issues and are meaningful for citizens there." There is not a single message, he asserted. "Modern dance isn't propaganda, it doesn't have a set form." Instead, the pieces are "open to interpretation." Hence the modernist neutrality of modern dance is also invoked as a means of allowing multiple interpretations, making arguments for one political ideology over another, or evading the recognition of content at all. As a modern form, it traverses borders, seemingly stands outside history, and allows claims to higher orders of abstraction to be made.

This exchange echoed themes of MMMs that took place several years earlier after performances by a Mexican modern dance company and by the Beijing Modern Dance Company. The dance companies were presented with the support of a grant whose terms were to present contemporary performing artists from countries with a high level of immigration to Southern California in the preceding twenty-five years. Political and economic dynamics reflecting international relations in the context of globalization imbricated the performances and their planning (Peterson 2007). Both groups provided opportunities for audiences to identify meaning related to the national context of the respective companies. Contradanza, the Mexican company, ended its performance with the women topless. Required to wear sheer leotards until the final performance because of Grand Performances' concern for its audience of members of the "general public," the intention of nudity was nevertheless obvious. At that evening's MMM a high school–aged audience member asked, in Spanish, "Coming from a closed culture and talking about a taboo concept, what happens in bed—sexuality, homosexuality, bisexuality—have you been censored in any Latin cultures?" "Not yet," was the response. An older man projected his knowledge of Mexican art onto the performance, collapsing millennia by saying, "I saw Tomayo, Sigueieros, I saw the sculpture of the Mayas." To this, a dancer replied simply, "Interesting."

Audience interpretation is key for the emergence of meanings of what is performed onstage, for drawing identifications between performance and nationality. It is based in a combination of what is seen onstage, artists' explanations, program notes, and prior knowledge. While on the one hand there is flexibility in the process of interpretation of a nonverbal form such as modern dance, at performances of the Beijing Modern Dance Company, overlapping understandings of international politics, knowledge of China as a geopolitical formation, and common assumptions about the performance of identity contributed to a relatively narrow range of audience interpretations voiced publicly. Relying on and emerging from this common field of interpretation, choreographers use symbols that point to yet do not determine extra-performative meaning. Legible Chinese content was conveyed through costume and choreography such that throughout, Chinese identity was conveyed as a gloss of culture and politics. Ultimately one would take precedence in order to posit particular narratives of Chinese identity and its relation to a global order. The discussion of these interpretations moved from the recognition of difference and identifiable Chineseness to an aspiration for universality.

The second half of the Beijing Modern Dance Company's program featured *All River Red*, set to Stravinsky's *Rite of Spring*. The program notes, written by the dance company's publicist, were printed in Mandarin and English and prepared audience members for what they were about to see. The two texts presented remarkably distinct descriptions of the intentions of the piece for these two linguistic publics of Grand Performances' audience, who, addressed as such, were posited as having different interests regarding the meaning of the piece. According to the English program text,

> This dance reflects the development of contemporary dance in 20th century China—a period when all artists in the country were required to do either propaganda or ethnic works. It was against this conservative backdrop that many Chinese artists gave up their social status and even lives in order to create contemporary and individual expressive work. *All River Red* is a poetic interpretation through dance to show not so much a comfortable fusion of the East-meets-West, but a direct and violent confrontation between those adhering to tradition and those willing to rebel.

The Chinese text, on the other hand, read,

> During the first performance of Stravinsky's *Rite of Spring* in 1933 [*sic*; the correct date is 1913], it signified the dawn of a new era in music. *All River Red* is this group's homage to *The Rite of Spring*. It reflects the development of twentieth-century Chinese modern dance from tradition to the modern age. Along the way many Chinese artists have made selfless contributions, creating special and significant contemporary works of art. (translated by John Osburg)

The English text focused on the struggle of the individual against society, evoking Chinese political history. It assured those reading English that at least some Chinese resist the oppression and conservatism of Chinese society, even if that resistance results in a "direct and violent confrontation" between East and West, tradition and modernity. This position could be read in the choreography in moments that appeared to convey narratives of exclusion, restriction, or struggle. For the Chinese-language audience, on the other hand, the company located itself in the history of European modern dance. In creating its own version of *The Rite of Spring*, the Beijing

Modern Dance Company entered the long history of realizations of the piece, from Nijinsky in 1913 to Lester Horton in 1937, Mary Wigman in 1957, and Pina Bausch in 1975. In participating in this history, the Beijing Modern Dance Company could become a member of the international (or Euro-American) modern dance world, deployed as part of a narrative of Chinese progress that moves from Eastern "tradition" to a "contemporary" West. Chinese audience members would also likely know that the title *All River Red*, which evokes Chinese communism to Americans, comes from a sixteenth-century Chinese poem arguably written by a hero of the Song Dynasty, the title of which is generally translated as *Full River Red*. It is taught to all Chinese schoolchildren to impart a message of the importance of rising to a challenge despite potential danger.[3] For both, the meanings of the program notes were sustained through what the audience saw onstage.

The performance begins. The bassoon solo that marks the "Introduction" of Stravinsky's *Rite of Spring* is heard from the speakers directed toward the audience. Dancers are standing in a group onstage, bathed in red light. All are facing stage right except a single male dancer, who, lit in blue, is turned away from the group. He holds up one arm. The group of dancers turns and runs to the man. One woman, left alone, is now in blue light. She holds a red cloth over a raised arm. As she lowers her arm she turns and moves toward the group, and everyone mingles. The dancers are wearing nude-colored leotards and hold red bags over their shoulders. They move slowly, bending slightly from their waists. One man starts to move with more animation. The company now stands in an organized group; the dancers all raise one arm and move more wildly. Three men leap through the air. The other dancers bend forward from their waists and move into a close group. The woman who has remained takes a dress from her bag and puts it on. The others follow her lead. The men wear Sun Yat-sen suits and the women wear red qipao with long slits up the side.

The use of red scarves as costumes and props reflects an intersection of a history of dance in China with Chinese politics and culture. Red banners are central to the history of dance in twentieth-century China. "Ubiquitous in revolutionary ballet," often appearing with portraits of Mao on them, "the connection of red banners with revolutionary women pre-dates the association of the red flag with Chinese communism—red banners or red kerchiefs were the emblem of the women's division of the revolutionary Boxers before and during their rebellion in 1900" (Strauss 1977:41). As the director (who is also director of the Guangdong Modern Dance Company) explained, describing some of the "Chinese" content of the piece, "Red handkerchiefs, the symbolic uses, are all ancient China; the handkerchief covering the face is a wedding scene. The color red is very Chinese, not necessarily communist."

The gendered clothing presents an opposition between tradition and modernity through the historic lens of the Cultural Revolution, the details of which audience members might have varying knowledge. The qipao emerged in the 1920s as distinctly Chinese clothing. During the Cultural Revolution it was seen as a symbol of capitalism. It was worn again in the late 1970s as China opened dialogue with the rest of the world and is now popular for weddings. Conversely, Sun Yat-sen suits were worn during the Cultural Revolution and continue to be worn as formal wear (Finnane 2008). The gray of the men's suits was allowed during the Cultural Revolution, while red, the traditional color for weddings, was banned. The setting of a wedding allows an interpretation of the clothing style and color as at once traditional and rebellious. The wedding moves between repression and resistance, as the context for reading the color red as representing Chinese culture and a politicized space in which red was not allowed to be worn for many years. Insofar as the wedding appears as a symbol of repression, rebelling against "marriage" might symbolize rebellion against "Chinese culture" in general. However, that capitalism and communism are represented by the respective garb of female and male, who in the end throw off their binds together, suggests a reading of the repression as that of the Cultural Revolution.

Invoking the sacrifice of the maiden that was central to the original *Rite of Spring* located the Beijing Modern Dance Company in the "development of twentieth-century Chinese modern dance from tradition to the modern age," as described in the Mandarin program notes. However, the dance company avoided a direct reproduction of the original Nijinsky choreography of the 1913 premiere that has been followed by many since, as the binding did not come at the end of the piece during "Sacrificial Dance." Moreover, binding rather than killing the woman appeared to be a metaphor for Chinese repression, of being "bound by recognition" in China, where, as the director explained, the state has only allowed dancers to do either folk dance or government propaganda. Instead of a "Sacrificial Dance" in which the maiden is killed at the end of the piece, the group as a whole is bound. Yet while the bound individual is carried offstage, the group prevails in its struggle; it overcomes its bonds, throwing them off in a collective act as an expression of freedom.

Combining the two narratives presented by the program notes, in a post-performance discussion the director elaborated how "East" and "West" can come together through the *Rite of Spring*. Invoking both the progress narrative of the Chinese program notes and the narrative of struggle of the English, he said, "I think it goes very well with the music. *Rite of Spring*, when first shown in 1913 in Paris, caused a riot. I think this is a Chinese response to the music; it's appropriate because we're basing it on our own culture." The infamous riot at the premiere of the *Rite of Spring* could be related to both texts, as it reflected the entry of the Beijing Modern Dance Company into a European dance history at the same time as it was an example of the transformative power of modern dance, which might come only with struggle against the status quo. The riot is often used to exemplify a mythical past when European classical music (or "music" in general) was "social" because there was a visible response rather than silent listening. Yet as others have shown, the response, in which audience members yelled at the performers and fought among themselves, some loving the performance and others hating it, was part of a complex intersection of an artistic avant-garde intending to shock, an audience interested in and accustomed to a different kind of music and dance, and vocal expression inside the theater as an appropriate means of voicing one's opinion (Kelly 2000). As Cocteau commented after seeing the performance, "the audience played the role it had to play; it immediately rebelled. It laughed, scoffed, whistled, cat-called, and perhaps might have got tired in the long run if the mob of the esthetes and a few musicians in their excessive zeal had not insulted and even jostled the people in the boxes. The uproar degenerated into a free-for-all" (qtd. in Kelly 2000:292).

The original *Rite of Spring* was part of an avant-garde modernism for which primitivism was one means of finding new forms of artistic expression (Huyssen 1986). *Rite of Spring* has been taken up to support various arguments. Adorno criticized Stravinsky's *Rite of Spring* for, among other things, its reactionary, rather than progressive, use of primitivism and its extra-musical content (1994:146). Stravinsky, however, in asserting that "music is, by its very nature, powerless to express anything at all, whether a feeling, an attitude of mind, a psychological mood, a phenomenon of nature" (qtd. in Peyser 1999:xvii), had already occluded the possibility that *The Rite of Spring* might convey a message. *The Rite of Spring* is also described as "the birth cry of modernism" and "the very prototype of the modern" (Acocella, Garafola, and Greene 1987:5). The Beijing Modern Dance Company draws on this latter interpretation, using it to reflect its own interest as at once situated within and overcoming "tradition" as part of a narrative of progress. Both elements are contained in the content of the piece. The use of symbols such as the red scarves that mark the performance as Chinese is a means of expressing Chinese culture and politics in modern dance while asserting that China is progressively becoming part of a European history by creating its own modern dance. In fact, the director said, "Five years ago *All River Red* could not be performed." But with the Olympics coming to Beijing, he continued, there is a need to show a different face of China to the world, one that is more progressive and open to artistic expression.

At later events in Grand Performance's season audience members would, without my solicitation, describe this performance as very beautiful and moving as well as legible in terms of narrative content. After seeing the performance, the dance critic for the *Los Angeles Times* related the visual iconography of the red flags to specific Chinese traditions, interpreting them as representing a "Maoist-style rally" and a "Chinese dragon dance." These "fabric-metaphors" supported an abstract narrative of progress, expressing "a powerful declaration of the need to cast away a heritage of mass brutality and embrace an entirely different kind of future" (Segal 2003). The legibility of a narrative of oppression stamping out future rebellions betrays outside knowledge of Chinese politics and traditions that are clearly relevant for the interpretation of the reviewer, suggesting that his understanding of the piece, like that of other audience members, is informed by information not provided in either the dance itself or the program notes.

The difference between the English and Mandarin program notes appeared to affect interpretations of the piece along national and ethnic lines. Following the now unspoken demands of the MMMs for audience members to ask questions that reflect differences of background, white American and Asian American audience members asked strikingly different questions. Several white audience members read the dance as a struggle against the communist government of China. The company director was asked to comment on this at two separate question-and-answer sessions. At one, a man said, "The last dance was obviously an allegory of some sort, I assume of recent Chinese history to some extent, Cultural Revolution,

Tiananmen Square. I wonder if you could be as specific as you can about what it was referring to."

The director responded that the dance was not about the government. "It's not a political statement against the regime. It's about culture, about the conservatism that artists have to face." In China, he continued, there is pressure as a dancer to be a folk dancer or to make government propaganda. There is little support for the artist who wishes to explore modern forms. Indeed, he explained, echoing the program notes, there was such repression during the Cultural Revolution that the only modern dancer in China at the time was jailed and others committed suicide rather than comply with these demands. So, he said, "If you are Chinese you must do a certain kind of movement, but of course when we are in a new era we would like to be treated as a general human being, not just as a stereotyped person. Of course," he continued, "it's dangerous. With Red China, when the communists came in, no wonder it gives you a feeling of the political, because maybe it is. Nowadays we cannot distinguish." In this instance the tension between culture and politics resulted in politics being neutralized as culture in order to express a marketable, serialized national identity without criticizing a government known for censoring such sentiments.

An Asian American audience member was less interested in the struggle of the artist against Chinese society and instead asked, "How does your dance troupe of Beijing promote the image of modern-day China? And how will it contribute toward China's contribution to world culture?" To this question the director responded that they were not trying to show an image of China. Rather,

> Always for me modern dance is about personal feelings, about being human, and if we show something truly ourselves then we have the chance to show the world what we truly are. Then we are successful. Because what we do onstage is from China, that will speak loud and clear about what the Chinese are becoming and what China is becoming. I hope it will do something good for China. Maybe the image is of suffering or of struggle, but at least there's freedom of expression allowed nowadays, so I think it's important that we truly do what we feel instead of trying to make a superficial image that probably may sell very well but isn't artistically sound.

The differing questions reflected the general tension between difference and universality that ran throughout the presentations and performances. The white American audience member drew on his knowledge of history and politics as a way of describing a fixed Chinese identity that is distinct and geographically remote and that necessarily entails a performance of something drawn from that identity. The Asian American, on the other hand, emphasized a changing China that interacts with and is connected to the rest of the world. In responding to the white American's question, the director described the piece as a response to a general and vague "Chinese culture" rather than the specificity of the Chinese government. The displacement of politics to culture allowed for a transcendence to a "general human" that a more specific narrative about life with the Chinese government might not, as the latter would embed the dancers in the particularity of the politics of the nation. In his response to the Asian American, the general human emerged as the "personal," which, as individual expression, could by extension also point to a new China. In asserting that the dancers embody a general "Chineseness" in an effort to be seen as "universal," to make art rather than sell a stereotype of China, the director repressed other possible narrative content of the piece. Instead, the dance came to reflect a new China that allows the liberal democratic ideal of freedom of expression, thereby promoting progress toward becoming a "general human being," a liberal subject in a neoliberal era.

Both perspectives contributed to the ways in which *All River Red* fit the goals of Grand Performances: a modern dance set to Stravinsky's *Rite of Spring*, a modern European composition, presenting a narrative—or multiple narratives—about China. Initially, Grand Performances' artistic director had wanted to present a piece set to Pink Floyd's *The Wall* that she had seen the group perform in Beijing, which used English popular music while evoking at once a Chinese monument and artistic struggle. However, the dance company was holding that piece for an upcoming European and American tour, and offered *All River Red* in its place. *All River Red* satisfied her desire to show an American audience something about Chinese culture in a form that was accessible to them. The ability of the audience members to understand the message of the piece—even if (or particularly) from their own perspective—was indicated by their questions about it. The presenters' aim to offer "a contemporary art form with traditional elements" was supported at the same time as the meaning of this expression was opened, taking on multiple valences through practice.

The success of identification that links dance or music and identity lies in the potential for multiple meanings to emerge in interpretation, meanings that are nevertheless constructed in a wider frame based on movement, text, prior knowledge, assumptions about the intentionality of the performance and the potential for representation, and ensuing conversations about these meanings that focus the possible interpretations. As a consideration of Chinese identity emerged in performance and post-performance discussions, a world beyond the borders of the nation was brought to California Plaza through the bodies of Chinese dancers and interpretations of audience members, sounding an Angeleno civic space that was at once of the city and the world.

Conclusion: Excluding Dissent

Sustaining the space of consensus that serves as the mode of public interaction and normative social dynamic of the civic as constituted by public concerts, performances that are explicitly political are generally excluded for their divisive potential. While such performances reflect the inextricability of politics and performance, a tension between the terms emerges in performances that are explicitly political or that articulate political modes of representation through art. Performances of East Los Angeles groups such as Culture Clash, El Vez, and Quetzal often blur the lines between music as harmonious and universal on the one hand and as a site for political action on the other. These groups self-consciously come out of and articulate a longer history of Chicano politics. Politics are expressed through the lyrics or narrative of a play, or a performance might be combined with an extra-artistic political or civic activity. Culture Clash and El Vez convey Chicano politics through humor and are popular across categories of identification.[4] Quetzal performed during Grand Performances' 1998 season as part of the Southwest Voter Registration Education Project Latino Vote '98 Conference. As the program notes stated, "SVREP is a national organization dedicated to furthering the political empowerment of Latinos and other minorities. We applaud their commitment to encouraging Americans to access and participate in the democratic process." Supporting political participation through voting sustains public concerts' role as civic institutions. Cultural citizenship is transformed into legal citizenship, as audience members become voters. Yet the political is limited to formal modes of democratic participation, as putting content to that vote would undermine the spirit of consensus promoted by public concerts.

Putting pressure on that consensus, Quetzal criticized President George

Bush from stage during its 2002 California Plaza performance. Many audience members applauded, but an older man came to the information booth and complained about the statement. "I want to enter a formal complaint against this group," he said. "Did you hear what they just said? They should know better than to disrespect our elected leaders. Everyone here should have booed them. They can't do that in a public venue." Grand Performances' staff responded by telling the audience member that this was a public place and hence allowed free speech for artists. While not taken seriously by the staff, this man's complaint indicates the pressure put on dissent and disagreement at free concerts. The political, his critique suggested, is divisive and as such should not be expressed in a civic space of consensus.

At civic performances particular identities are acknowledged and supported as modes of belonging and ultimately transcended in the service of recognition of common membership. Processes of identification reveal their own making, as negotiations over interpretation of musical signification bring together shifting fields of culture and politics, national and international identities, exposing the particularity of the general public in its making. The figure of the "universal human" draws a world to Grand Performances, making Los Angeles's civic space a global stage in a modern order. With modernism circulating as an international trope, consensus of interpretation at these performances becomes consensus on the level of the world. And though a link between music and identity enables public concerts to be sites of belonging, supporting a space of consensus that is built on aspirations of social inclusion requires the exclusion of dissent.

5

"Los Angeles at Its Best"

Plant a stake crowned with flowers in the middle of a
square; gather the people together there, and you will have
a festival. Do better yet; let the spectators become an
entertainment to themselves; make them actors
themselves; do it so that each sees and loves himself in the
others so that all will be better united.

—Jean-Jacques Rousseau

Los Angeles is at its best at Grand Performances. This is a
diverse place, but usually everyone is separate. Here
people come together—people of all different
socioeconomic groups and races.

—Grand Performances audience member

"So the next song, this is a very special song, I'm going to try to hold it
together emotionally, as we play this, it's one of them songs you know, one
of them songs. . . . It was written by Parliament and Funkadelic." Hearing
this, some in the audience yell in appreciation and recognition. "It's called
'Come In out of the Rain.'" Double G cues the orchestra and we start
playing. A man in the audience who was, as he described it later on his
Web site, already in "almost full groove" was overjoyed at hearing the song
announced, his personal dedication to Parliament Funkadelic emblazoned
on his sweatshirt with the letters PFUNK (PFUNK 2003). People dance in
the Upper Plaza. One person waves her arms in the air, moving to the
music. Another, with a bleached crew cut reminiscent of white rapper

Eminem's hairstyle, beats his head, a little off the beat, pauses, finds the beat, and then is on his own time again. The solo violinist starts the guitar solo of the original Parliament recording. She plays wildly, her left hand at the farthest edge of the fingerboard and her face contorting with expression. Her long brown curls fly and she bends backward as she reaches the highest point, repeating the notes over and over, their sound given a warbly twang by the fuzz/wah pedal that she presses with her foot. The lights illuminating the musicians flash like strobes and the MCs sing the chorus, "When will the people start gettin' together, learning to live and love one another, he-e-e-y hey, h-e-e-e-y hey. When will the people start findin' each other, learnin' to give and help one another, he-e-e-y hey, he-e-e-y hey."

Written in the early 1970s as a critique of the violence of the U.S. involvement in the Vietnam War and revolutionary protest tactics in urban America, when performed at California Plaza in 2003, the lyrics of "Come In out of the Rain" articulate aspirations for an experience at public concerts made possible by musical performance. Peace as a political position becomes a sensibility for civic participation organized around consensus. Rather than taking a stance on urban and international social concerns, the audience looks to itself for the recognition of social aspirations, allowing it to serve as a synecdoche of the city. At civic performances, music is a motor for the realization of ideals of public life. In a process that is at once sonic and spatial, music brings people together, sounding an ideal Los Angeles that is heard, felt, and danced. These experiences evoke the recognition and fulfillment of some of the general aspirations for Grand Performances, for Los Angeles, and for contemporary public life.

An audience is constituted through affective, participatory, sensory experience. Depending on spatial proximity, it is "a crowd witnessing itself in visible space" (Warner 2002:66). Yet the live audience remains largely untheorized. What is this face-to-face group? What meanings are generated from the experience of audience members? And how, as Barber asks, "do audiences do their work of interpretation" (1997:357)? What are they interpreting and to what effect? How does a moral community emerge from affective experiences of individuals who listen, dance, and watch performers onstage? The affective, embodied, experiential, and sensory recede in discussions of text-based publics, in which moments of coming together are primarily understood to validate identities generated by consumption of circulating media (Warner 2002).[1] Yet public concerts suggest something significant is happening with audiences of live events and that what is happening there might reveal critical insight into the contemporary condition. The public concert audience is constituted and recognized as a moral community that is in and of the city, fulfilling urban ideals through embodied experiences of listening and dancing together. The particular modes by which a moral community is created at public concerts indicate the nature of an audience of live events and its potential for generating meaning beyond the event itself. However, while yielding an understanding of audiences generally, public concert audiences are specifically civic collectivities, insofar as they invoke a recognition of the city and of an urban public that is captured "at its best" in the public concert audience.

The simultaneity of performance facilitates a collective experience that, sounding an intimate and emotional space, in turn generates broader social meanings for the audience. Aspirations for togetherness maintained as the potential of public space are realized and validated through the experiences created by the public concerts (Low, Taplin, and Scheld 2005; Whyte 2001). In feeling a sense of togetherness and community, an affective public is produced that depends on physical proximity and recognizes that fact as central to the efficacy of the public concert. While the value attributed to "liveness" is in part a response to contemporary forms of mediation, that value is itself significant for what it tells us about the contemporary condition and aspirations for what it might be. In and of the city, sound and space support an experience that is measured against a segregated, dispersed city.

Civic performance helps realize utopic aspirations for Los Angeles's diversity by bringing people from a segregated city together downtown. There, feelings of cohesion and community generate social meaning and a recognition of a diverse L.A. The success of this narrative is heightened with dancing, which facilitates, through a shared ideological frame of ritual, the movement from self to collective, individual to community, dancing body to Los Angeles. Audience commonality and unification are accomplished in the face of divergent positions and practices, as embodied experience closes the gaps, authenticating a dominant narrative and ideal of a diverse city. Not all performances showcase dance music, and even at those that do, not everyone dances. Yet in describing their experience of dancing and its importance for the meaning of public concerts, those who dance or find meaning in others dancing define the forms of legitimacy of the event and write the terms of consensus for public concerts. For presenters, the performances at which people dance are the moments when the inexpressible "it" happens that reflects the best version of Grand Performances.

"Gettin' Together"

Midway through daKAH's concert, more people in the audience are dancing than before. In the Marina Pavilion—an area designated by staff and claimed by audience members as a place to dance—people in the audience stand, their bodies directed toward the stage, their eyes on the performers. They move like a sea of grass, each individual bending in his or her own direction, feet rooted on the ground, in time to a common beat. One man, more energetic, bounces up and down, while a woman in a white skirt and tank top moves her feet and runs her hands through her hair. At the edge of the Marina Pavilion, next to the chairs in the Lower Plaza, people sit on the ground and bounce to the music with their shoulders and heads. They watch the MCs intently. At the back of the stage, behind the shooters of the fountain, people stand, talk to friends, and put their hands in the water. A middle-aged Asian couple walks by, man and woman both slightly hunched over, with short hair and glasses, wearing baggy khaki jackets. They look briefly at a skinny man wearing a tie-dyed T-shirt tucked into his khaki pants dancing alone behind the fountain. He is at the outer edge of the Plaza, close to a row of trees that marks a border between the performance area and the courtyard of the Omni Hotel. People walk past him in a steady stream. Another man, wearing a dark sweatshirt with a white T-shirt peeking out at the bottom and holding a backpack by its top hook, dances next to him. They look at each other briefly, acknowledging that they are dancing together.

After the concert PFUNK writes about his experience of dancing with a stranger in the audience, posting it on his Web site under "subjective experience" (PFUNK 2003). PFUNK's narrative starts with his recognition of a man dancing next to him. Their location at the front of the Marina Pavilion puts them close to the performers while also making them visible to the audience around them, such that their personal experience was also a performance for others in the audience of a "big ole black 40 yr old american male" dancing with "a 50 sum'n polynesian guy." The "common beat" of the music enabled them to feel the groove and dance together. Dancing allows for nonverbal communication between individuals; in the feeling of togetherness, of moving to the same beat and knowing they are moving together, a moral community is formed between two that can extend to the whole. This experience "informs individuals that they are in harmony and makes them conscious of their moral unity. It is by uttering the same cry, pronouncing the same word, or performing the same gesture in regard to some object that they become and feel themselves to be in unison" (Durkheim 1915:262).

PFUNK continues with a description of the slow process by which this communication and recognition took place, explaining how it took him some time to find the other man's rhythm.

> we was gettin it though we never acknowledged each other and to tell the truth his eyes could have been closed but even so he heard my finger snaps cause they were right there as i held it high for everyone to see as i smiled with enjoyment from the feeling i was getting. i was publicly enjoying a good feeling and enjoying the fact that i was enjoying it in public. almost a self perpetuating energy.

The performance of dancing was integral to PFUNK's personal experience: seeing himself recognized by others reinforced his enjoyment in dancing to the beat with a stranger. His pleasure, he felt, would be infectious, spreading to those who might be watching him. The music onstage created the conditions for this experience, providing the "common beat" to which two men could dance—even in very different styles—without speaking or otherwise obviously registering their recognition of each other. The lyrics, "When will the people start gettin' together, learning to live and love one another," are echoed in the experience of the audience members. The affective public of the audience does not depend on verbal communication but can emerge through physicality and presence that is felt as something shared even if the other individual involved is not actively recognized with a meeting of eyes or a word.

Articulating a theory of ritual, the audience member's body is the site from which subjective experience generates a collective (Bell 1992; Comaroff 1985; Durkheim 1915; Mitchell 1956). According to Bell, "ritual theorists, experts, and participants are pulled into a complex circle of interdependence" (1997:265), such that the concept of ritual has influenced how Europeans and Americans "ritualize" (1997:262–63).[2] Following Durkheim, utopic, ideal versions of society are drawn from and are necessary for the formation of the social. Like Durkheim's "ideal society," the public concert "is not outside of the real society; it is a part of it." Moreover, "society has constructed this new world in constructing itself, since it is society which this expresses" (1915:470, 471). The "Los Angeles at its best" recognized at the public concert follows Durkheim's social theory; as utopic aspiration, it comes out of an understanding and experience of the social totality limited, here, to Los Angeles. The totem, or representation of the symbol of the whole, is, in this case, the audience, which—recognized as symbolizing diversity—serves as a representation of the multicultural city.

Ultimately the audience turns toward itself to give meaning to the event, interpreting the concert as successful by performing itself as a collective. Following Rousseau, "The spectators become an entertainment to themselves," and "each sees and loves himself in the others so that all" are "better united" (1960:126). Today, in Los Angeles, Rousseau's call for civic performance in Geneva serves as an aspiration for a consenting public that might experience a collective "togetherness" against fragmentation that is both projected and experienced and a public that is recognized as representing a diverse city. Universalizing tendencies of ritual, conditioned by Durkheim, who opened up "what would be an ahistorical, sociological approach to religion as a functioning system of social relations" (Bell 1997:25), are deployed even as they are made specific to the contemporary city, to Los Angeles figured at once as fact and fantasy.

Durkheim framed *The Elementary Forms of Religious Life* as an aspiration, the examples pointing to kinds of experiences that might offer an alternative to what he considered the anomie of the contemporary condition. According to Durkheim, aspirations for an ideal society are generated from experience and meaning making, or, in his words, rite and belief. In order to sustain the beliefs that support the social, rite and belief must be continually engaged, performed and re-performed. The feelings generated by group experience, "impressions of joy, of interior peace, of serenity, of enthusiasm," are, "for the believer, an experimental proof of his beliefs" (1915:464). At civic performances, the togetherness that stands against urban sprawl, segregation, and a city of neighborhoods is recognized in a unified audience produced as such through sound engineering, performance techniques, and audience experience and recognition of dancing and diversity. While these processes contribute differently to constituting the audience as a collective, all serve to create the conditions for affective, communal experiences that can be recognized as such. Such recognition might be personal feeling, visual acknowledgment of others, or aural expression;

while a range of experiences exist at once, as long as someone articulates or performs the required participation the potential for recognition of the ideals is possible.

The audience becomes a collective through sonic intimacy and spatial proximity. Drawing belief from rite, audience members discuss their experience in terms of the tropes of ritual: outside everyday life, collective vocal and bodily expressions that become "songs and dances" (Durkheim 1915:247), regular rhythms that help participants experience a feeling of togetherness, and temporally concentrated events. Hearing, feeling the beat, and dancing are ways in which audience members feel connected to each other, allowing them to become part of a collective. Echoing PFUNK's experience, another audience member commented that a public concert is "one of those places you can get together with all kinds of different people and there's that wonderful group energy where you're just enjoying and grooving on something together." Continuing, she said, "music in particular just has this ability to create this sense of unity and oneness in people that I just think is so magical. It's sort of like the universal pleasure machine."

She projects a sense of togetherness from her body as she dances, finding therein a connection to others that is facilitated by hearing the same sound and being in the same space. Marking its significance for the contemporary moment, she maintained, "we sure do need it these days. It's one of those few places where, no matter what your politics is, if it's a great beat everyone's smiling and moving and happy together and it just kind of reminds you that we're one big community." For this audience member, politics is removed, difference drops away, and community is found in one woman's experience, in her dancing body. Community, in which there is a sense of "oneness," requires the removal of difference such that all are the same, dancing together and feeling happy. A self-validating process, she knows from her own experience that this is how she feels. She includes the rest of the audience ("everyone") in her feeling, finding evidence for the experience of others in the fact they are "smiling and moving."

While the recognition of the collective in the individual articulated by public concert audience members, performers, and staff invokes a Durkheimian theory of ritual, this is not the ritual of "'untheorized' collective action" (Comaroff and Comaroff 1992:71). In describing their experiences in the language and model of ritual theory, Grand Performances' audience members articulate theories of liberalism; finding features of ritual ascribed by social scientists primarily to non-Euro-American societies in their experience of attending public concerts, they use a Durkheimian framework to express their understandings of how public concerts fulfill ideals of urban public life. My use of Durkheim to analyze these processes, therefore, is a way of drawing on "local theory" while using its status as social theory to develop an analysis of how public concerts create meaning, of how a public concert audience is recognized as reflecting a diverse Los Angeles. The purpose in engaging with this social theory is to consider what it might offer for understanding a present meaning: the particular content that is apparent in the experience of public concert audiences that tells us about the significance of public concerts for Los Angeles today. Invoking a universalizing framework of ritual is one of the ways by which public concerts acquire meaning, allowing them to transcend the space of the corporate plaza, the performance of the taiko drummer, and the everyday lives of the audience members. A framework of ritual naturalizes experience, as past ideals are recuperated for present purposes. Helping produce a consensus around the dominant paradigm of diversity, experience writ as myth serves to occlude both the conditions of possibility for that experience and other potential characteristics of the city.

Sound produces experiences of spatial proximity that are felt first in the body. Bodies feel, becoming sites from which affective meaning is generated. The body is a privileged space where audience members locate senses and affect, or nonverbal ways of processing and making meaning. Locating these ideals in the dancing body of one's self and of others—a process in which "embodiment" is coded as more real or true than language—contributes to the efficacy of concerts as sites for community and cross-cultural understanding (Merleau-Ponty 2002). Music is considered unique in its ability to reach the nondiscursive body because, as an audience member explained, it "directly passes the left brain and goes straight to the right brain in emotion. Most of people's prejudices are based on things they've been taught or read or believe and music kind of takes you to another level. It has an incredible power to move people emotionally and to move them in beautiful ways." As an experience located in one's own body becomes a means to directly connect to a collectivity, particular cognitive or discursive prejudices are overcome in this space of feeling and affect, a "level" where individual differences are put aside and commonality is felt: "Once [meanings] have taken root in the body and have acquired a natural alibi . . . these meanings take on the appearance of transcendent truths" (Comaroff and Comaroff 1992:71). In this way, the "universal" nature of embodied experience helps naturalize an ideal of a diverse Los Angeles.

The body also articulates myriad social meanings, embodying, as it were, categories of race, ethnicity, gender, class, age, ability and disability, and normativity and the abnormal. Insofar as the individual body is always also produced as a historical subject as both individual and body, it is at once a subject of and productive of power (Foucault 1977, 1978). Locating the social in the body of the individual evokes this field of power—which in this case is primarily the ideologies of a diverse Los Angeles—that, in helping produce the individual body as such, enables it to serve as the locus of these ideals and aspirations. Bodies at public concerts are both a means of experiencing something construed as universal at the same time as they are identified as marked and distinct, the material of a diverse audience and city. The body, with its ability to express social ideals through experience, is produced as such through processes of recognition of its own meaning and the embodiment of other meanings. In "performing" wider social processes—that already exist as ideals of the very events at which they are found—dancing at public concerts serves as "a reiteration of a norm or set of norms" (Butler 1993:12), a way of recognizing and generating affect, community, and diversity.

Civic Senses

Music is embodied as it is heard by the ear. When loud, music, in particular the bass register, can be felt throughout the body. As the source of the impetus to dance, music is imbued in and brings an awareness of the physicality of the dancing body. A public concert audience is unified by hearing the same sound from the stage. At California Plaza, the audience hears an amplified projection of what is played onstage. Sound, space, and social life coalesce as sound technologies—such as the locations of the speakers, the distance the sound projects, and the volume of the sound—support the experiences of those within the Plaza, draw social boundaries, and mark the terms of normative urban civility. The speakers that relay the music to the audience are carefully placed so that people in different areas of the Plaza will feel they are hearing the sounds at the same time. They are also angled so the sound from one does not overlap with that of another. Rather than creating a seamless transition from one speaker's area to the next, a slight gap in sound is created; thus a person moving around the Plaza or standing in the area between two speakers has the impression of hearing a continuous sound rather than overlapping moments of the music caused by the time delay between the two speakers. The speakers are discreetly located in order to convey the impression they are not, in fact, there. They are hung in scaffolding at the front of the stage, in the scaffolding over the front of the Marina Pavilion, and under the overhang on the Lower Plaza.

In creating a seemingly unified listening experience that constitutes the audience as a group, amplified sound produces multiple sociabilities. The gaps between the speakers create distinctions among the audience as people in different areas hear slightly different moments of the music. These gaps are re-closed through the technical control of sound as a metaphor of social amelioration. This smoothing over of social differences is reproduced through volume, the levels and control of which imbue and reflect wider social normativities. The technical director admits that the best sound is right in the middle, where the soundboard is located. Thus the technologically mediated sound posits the perfect audience member as one with a centered body and sophisticated ear; the audience member with these qualities is granted the perfect listening experience and becomes the perfect listener who hears the best and most balanced sound.

Municipal noise ordinances that maintain the regulation of sound in public inscribe sociabilities through rules about space, time, and volume (Bijsterveld 2008; Thompson 2002). Prescribing normative categories of sound, music, and noise that have long been the subject of debate (Attali 1999; Cage 1973), the Los Angeles Municipal Code allows amplified sounds of only "music and/or human speech" (115.02[e]). Such normal and expected sounds in turn posit the normal subject through their proper performance, which should be "limited in volume, tone and intensity" (115.02[f]) so as to not be "loud and raucous or unreasonably jarring, disturbing, annoying or a nuisance to reasonable *persons of normal sensitiveness*" (115.02[f.2]; italics added). A person's normality, structured according to sensitivity to sound, reflects wider social relations. As Bourdieu maintains, "thresholds of tolerance to . . . noise" are manifestations of "a lasting relation to the world and to others" (1984:77). Extended from the person to the city, permissible decibel levels systematize urban social codes through sound technologies.

Tensions of liberalism are inscribed in noise ordinances, pitting public against private through zoning regulations that vary according to commercial and noncommercial sound production, allowing less sound near

Figure 10. A speaker amplifying the acoustic sound of an afternoon performance. Photo by author.

churches, schools, hospitals, and residential areas. Oppositions between public good and private property are thus produced through sound. Legislating that sound not be heard beyond the "enclosure" or "property line" of the space from which it is made (IV:41.32), the law invokes the enclosure movement of eighteenth-century England that marked modern capitalism's early spread into the social fabric (Polanyi 2001). In the name of the "public health, comfort, safety, and welfare of its citizenry" (XI:5:115.01), the interests of the abstract general public privilege "freedom from" unwanted sound over "freedom to" speak or make sound in ways that might be considered aspects of democratic expression. Sound is thus ordered under individual, private experience, mandating that being in public is inscribed within the boundaries of the private, in terms of both property and experience.

The "normal person" who is produced through enforcement and complaint is supported by noise regulations that are nevertheless socially and subjectively enforced. Called by disturbed neighbors, police are encouraged to use personal discretion and subjective determinations in the enforcement of noise ordinances (Los Angeles Police Department 2004–8). Hence this law that structures norms of urban civility through sound production and regulation supports the use of sound to maintain social controls and prejudices. Only rarely do audience members complain about the sound in California Plaza, and if they do it is usually that the volume is too high. The manager of One California Plaza said that the sound increases in volume as it rises, and during lunch-hour concerts he would receive phone calls from workers in the upper floors of the office towers complaining that the sound from noontime performances was affecting their ability to work. Such complaints position the interests of residents and workers over those of audience members, older people over younger, private over public, work and home over entertainment and leisure.

The Los Angeles Municipal Code maintains that amplified music can only be played between 9 A.M. and 4:30 P.M. in or within 500 feet of a residential zone. The times during which sound is allowed to seep outside its boundaries support labor over leisure and presume a "normal" working day. Given that Angelus Plaza is approximately 175 to 200 feet from the amphitheater, special precautions and subjective judgments must be made. In response to previous complaints, the staff tries to not let concerts go past 10:30 in the evening so that the senior residents of Angelus Plaza are not bothered. Though personally unaware of the specificities of the noise ordinance, the sound technicians are "neighborly"—as they put it—by angling the speakers so they are directed toward the audience and enforcing their own evening cutoff. As a result, there were few complaints about the noise from the residents over the first seventeen years of the series.

The possibility for dancing emerges within a context in which social status bears on volume, genre, and comportment. The sound technician adjusts the volume according to the type of music being performed. European classical music is played at the lowest volume. For classical music, the technical director places the microphones on the ground to get what he described as "air," which, he said, helps create the sense one is hearing the instrument directly. He places microphones behind violins to get what he calls a "warmer" sound. By hiding the microphones so they are not visible to the audience, audience members' ears will, he asserts, play tricks on them, making them think they are hearing acoustic sounds. Technology entails mediation between the body and the music that can be considered undesirable, diminishing the direct connection between self and sound. The years of experience of the technical staff enable them to create a sound at performances that can give the impression that it is not amplified. Sound engineers increase the volume and bass frequencies for rock music and other dance music. "You want the bass," the technical director said, "because you can feel it physically." When audience members feel the sound in their bodies, he asserts, they are more likely to dance and to have a good time. He described a noon performance of a popular East L.A. band during which people working in the office towers had complained about the volume. The audience was dancing, hands in the air, but when the sound technicians turned down the volume they "saw the hands go down" and people started to meander away. They lost the audience to the volume, he explained. The volume helps overcome the distance between the performer and the audience created by the water in the front of the stage.

For daKAH the volume is high. By the middle of the concert, many people are dancing. The drum continues steadily and strongly and the MCs are on the beat, words and sound effects tripping over their tongues. The cellists sit back and relax and continue repeating the section. We are not dancing in our seats now. We cannot hear the MCs and can barely see them at the front of the stage. The part we play is not technically difficult, but some energy must be put into playing it accurately time after time. Our part alone gets a bit boring. Maybe we are not in the right frame of mind for the repetition, the aesthetic of the drum machine. At the moment, this is not the pleasurable groove discussed by Keil and Feld, in which the musician feels "deep identification or participatory consciousness" as "you flow into repetition" (1994:23). Rather, it is the boredom classical cellists often express about playing the repeating two-bar progression of Pachelbel's "Canon" that continues, unchanging, for the entire sixty-four measures of the piece as the violins play variation after variation of the melody, which become increasingly technically difficult. Keil and Feld attribute this attitude of boredom toward a "Western fetish that novelty is progress and newness is what it's about" (1994:23). Yet the importance of the repetition is not lost in all "Western" music; for the dancers in the audience at this hip-hop concert, this groove of a repeating bass line or drum beat is essential to their ability to dance. The discrepancy between the feeling of the dancers and the feeling of the musicians is summed up by a statement made by a shakuhachi player, who explained how he could double the tempo of what he was playing and "make the audience go wild" without expending more energy himself. Thus, while the energy and focus of the performers matter for the audience experience, the performers might not be having the same experience as the audience. Moreover, performers within a group might experience the event differently from one another. Later Double G said this was the one time in the evening when he really felt it was working, when it had "that feeling." The MCs were having a creative, energetic moment, supported by our repetitions, inspiring him to keep repeating the section.

A dancing audience can be a sign of approval that marks a successful performance. Double G, hearing that people—at least those standing in the Marina Pavilion—had not danced much during the first half, said, "That's because we weren't getting them to dance." As a young rapper told journalist William Shaw, when a DJ played his song at a club one night he was ecstatic. He could tell they liked the song because "everyone was dancing. Like two, three hundred people. And they danced all the way through. All the way through" (Shaw 2002:325). Performers have different means of encouraging audiences to dance if they are not already doing so. Nati Cano, a mariachi musician who performed later that season, urged the audience to dance, saying, "If you dance, it shows we're successful." At Linda Tillery's concert, one of the singers started dancing. At one point she moved her arms, one after another, continuing her dance as she reached out toward the audience. As she finished the motion with her left arm, a whole section of the audience rose from their seats and started to dance. Performers might also speak to the audience in the course of a song. The energy is usually heightened by the increased volume of the voice of the speaker, who encourages interaction between the audience and him- or herself by getting the audience to yell words. At hip-hop performances this is often, "Say hey!" "Hey!" "Say hey!" "Hey!" The audience is "played," with one side of the room asked to compete against the other by producing the most volume. The MC can often get the audience to be louder by yelling more loudly through the microphone. A listless audience that is not paying attention to the performers can be engaged by having words directed toward it that also require its participation as more than passive listeners.

Passive listening, or what appears to be passive listening, is the norm. Most people at concerts sit quietly, with their bodies and eyes directed toward the performers. They clap at the end of pieces. This is the kind of engagement expected and required by this context of performance practice: the concert (or theater or dance performance) in early twenty-first-century United States. Audience expression appears illegible and private, as there is little outward facial or bodily expression visible to observers. Yet a dichotomy of audience as either spectator or participant mistakes convention for expression.[3] While "spectator" implies passivity and assent (Kruger 1992:4), the silent audience is never only a passive observer. Rather, the silent audience is a sociohistoric configuration. Silent, passive faces cover the inner, personal emotive response that we have learned from the European Romantic era is appropriate (Johnson 1995). The practice of the silent listener who makes sound at appropriate moments by applauding or yelling further marks the audience's consent, necessary for the legitimation of the concert form.

The listening experience of the silent audience, seemingly unmediated, marks a historical relation between expression and emotion that makes the latter illegible to anyone other than the individual having the experience. Performing this private response with others makes it a public activity. Silence acknowledges and shows respect for the public of which the listener is a part; speaking aloud to one's neighbor during a performance bothers other audience members, infringing on their private qua public enjoyment. Conversely, at comedic events, the whole audience laughs together, individuals unselfconscious as they express their response to the humor. However, when an audience member notices he is being observed, he often appears uncomfortable. Despite the fact that this is a public forum, the discomfort exhibited in these situations suggests that the participation and expression of the audience are thought, or desired, to be private. It is the performer who is on display for the public, not the audience. Sometimes the audience member makes a shift from private to public expression, such as PFUNK describing his dancing experience and his awareness of being watched. But even at public events dancing, too, can be experienced as a private, personal affair, as indicated by the dancing woman who, after being photographed, stuck her tongue out at the photographer.

At some point during daKAH's performance it appears, from where the cellists sit at the back of the stage, that no one is dancing and that the audience is not responding physically. From the stage, all we see are the people at the front of the audience: those who came early for seats and those who are standing and watching intently. We can see empty seats where a few people who had reserved seats did not show up. The people visible to us mostly sit still throughout the concert. It is difficult, especially with this music, danceable music usually played in clubs, to know whether people in the audience are enjoying the performance. This is compounded by the fact they are on the other side of a body of water. Many musicians who play at California Plaza complain about the distance between themselves and the audience. While this distance enables the possibility for the audience to focus on itself, it also means that greater effort must be put into looking at what is onstage. The connection with the audience that musicians feel when playing in other venues or even in the Marina Pavilion is lacking. Now the connection has to be made and experienced aurally rather than visually. What the musicians cannot see at all is the constant activity at the outer edge of the audience, just past the point where, from the audience, it is hard to see the musicians onstage over other audience members' heads. There people dance, and walk around and talk with each other.

Though we cannot see them dancing, we can hear people yelling and screaming and applauding. Screaming in recognition when a performer plays a familiar song at a live performance indicates that the person screaming (or clapping or smiling, depending on the genre) has heard the song before. The screaming is based on a personal relationship with the song; repetition evokes the experience of having heard it before. The song resonates with that individual personally, while at the same time screaming at the concert becomes a means of performing membership in the public of all those who have also heard that song, at that moment also recognizing each other as members of that public. Screaming, to be audible to the rest of the audience and the performers, conveys an expression of affective presence and connection across the water and through the granite plaza.

Dancing a Diverse City

Public concerts are heightened spaces for the recognition of general aspirations for Los Angeles: of celebrating its diversity while overcoming difference and of facilitating a sense of "community" that might support this. The public concert audience—as a collective and as community—is immediately glossed as diverse. Audience members evaluate the events as successfully creating a feeling of groupness and togetherness, validations that are followed closely by articulations of their role as spaces of diversity in which difference can be overcome. Public concerts "bring people together," creating "a sense of connectedness and community"; as such, an audience member said, they are "one of the few ways that all races, creeds, and colors can join together to enjoy artistic events."

The performance is instrumental to other social goods and ideals, insofar as "it encourages community from lots of walks of life" and "encourages people from all walks of life to share joy." Moreover, public concerts inspire moral transformations by bringing "community together to understand different cultures." Recognizing this experience, a presenter said, "I think people like to come here because they like to be with people different from themselves." The experience and its articulation in these terms must be performed and re-performed, with belief generated from and validating participation in the "rite." The production and validation of the significance of public concerts that comes out of these experiences is—as it must be—articulated widely, supporting both what is and what will be. Past experiences become motivation for future participation on the part of audience members and for presenters to organize a series that facilitates these experiences.

The audience "sells" Grand Performances. In discussing a fund-raiser, a board member explained that in order to raise money from people who do not already give to Grand Performances "it takes getting people here: once they see the audience" they will be committed to the organization. The board, which supports the organization materially and symbolically, places a keen emphasis on those performances that, in creating a heightened experience, allow for the realization of the ideals and aspirations of Grand Performances. Board members describe some of their initial encounters with Grand Performances through the experience of dancing or of being in the midst of others who were dancing. A staff member described how the first performance she went to after she had been hired was one at which, "At the end, in the Marina Pavilion, there were sexy, amazing dancers. On the other side, there was a grandfather with his granddaughter dancing, people of all ages, and I thought, I've come to the right place."

Visual recognition of diversity gains meaning in coming from the experience of dancing. At a retreat, board members drew pictures of their favorite memories of Grand Performances. These emphasized experiences of dancing, of being part of a group, and of the diversity of the audience. As visual representations, two stand out that, though different, show the tropes of audience as group, heightened experience, dancing, and diversity. In both, the audience is the focus of the meaning of Grand Performances as represented by the drawings. Portrayed as a group, one highlights the artist, dancing among others, conveying her experience of joy in dancing at daKAH's concert with a smile on her face and lines emanating from her head. The other, with repeated patterns of people and stars, has music represented with treble clefs but no musician; instead, the music is clearly instrumental to the audience, which, as similar stick figures drawn in different colors, is represented as diverse and as a collective. Visual representation mirrors the importance of vision in recognizing the diversity of the audience at performances. While dancing is first felt, skin color, as epidermal signifier, is seen. The movement from hearing and bodily affect to visual recognition creates a total sensory experience that is partially reproduced through the physical act of drawing pictures that seem to convey, most of all, the energy of the event.

Grand Performances programs world music, especially world beat, to create a multicultural audience and to encourage dancing at public concerts. The music used for the civic aims of public concerts is historically specific. At the end of the nineteenth century, brass bands—a popular trend of musical participation that swept the country—provided the music at many free concerts (Starr 1987). In the early twentieth century European classical music was played to immigrants at free concerts as part of a project of Americanizing them, as it was understood to both elevate and erase differences (Campbell 2000). At early twenty-first century civic performances, world music or world beat holds a special ability to both mark and "transcend" differences. World music supports a serial multiculturalism that, in bringing audiences of different backgrounds, facilitates the recogni-

Figure 11. "This is daKAH. It's about energy and joy." Courtesy of Grand Performances and the artist.

tion of diversity, of cross-cultural understanding, and of a community that transcends the spatial segregation of the city. Mixed into a summer that might include modern dance, European classical music, folk music, and hip-hop, events at which people will dance facilitate repetition of an experience that is the heightened moment from which understandings of togetherness are generated. World music has served the function of being a music to which people like to dance since its inception as a genre of popular music, when it emerged as "a global industry focused on marketing, managing, promoting and circulating danceable ethnicity on the world pleasure and commodity map" (Feld 2000:179). Depending on social codings and personal taste, other kinds of music might not achieve these ends of harmony and pleasure.

Articulating personal taste that nevertheless reflects broader patterns emblematic of contemporary social alignments with musical genre, one woman claimed that neither heavy metal, classical, nor jazz creates these

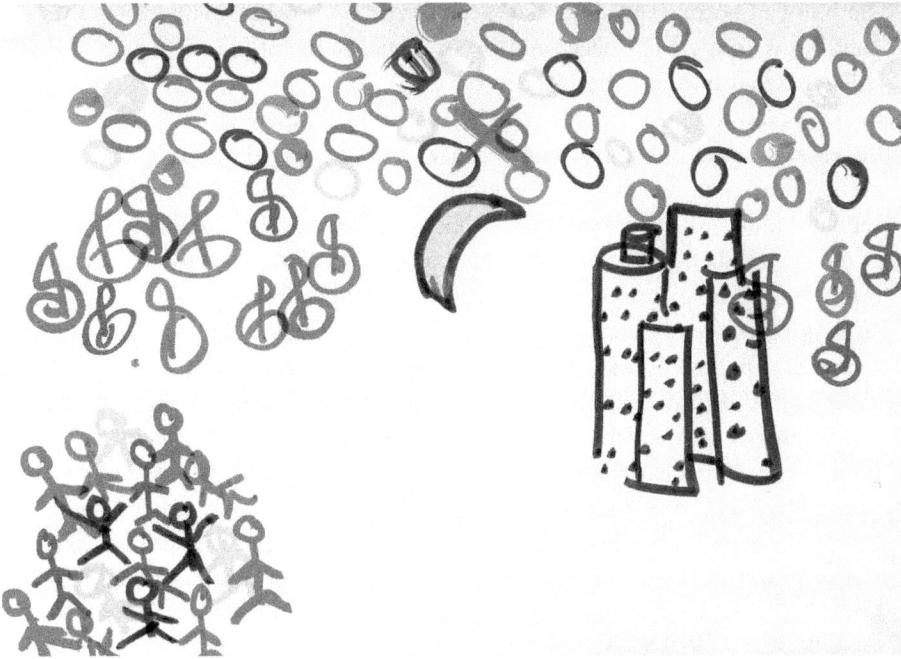

Figure 12. "This picture is exemplary of the diversity of people. I love the diversity of music and people dancing. I love being in the midst of that." Courtesy of Grand Performances and the artist.

conditions. As she explained, the kind of music that gives "an encompassing feeling" is "really great groovin' world beat stuff." This gave her the feeling it was more directly connected to her body than classical or jazz, which for her are more intellectual. On the other hand, she did not consider heavy metal universal because it was a means of expressing individual anger that would be directed against someone else or the world at large, thereby dividing rather than uniting. Rap, especially "rap music that's talking about killing cops" rather than contributing to commonality among listeners, "isolates" and "makes you angry about your circumstance." DaKAH, with its politically progressive lyrics and European classical instruments, overcame this barrier. She also excluded European classical music because it is not liked by everyone; ostensibly the exclusion is along generational lines, as articulated here, but its exclusions extend to its lack of danceability and relation to class. World beat, in being danceable and differently marked to the point of becoming unmarked, generalized as "world," helps create the conditions and feelings necessary for an audience member to feel part of a collective that in turn reflects a diverse city.

In describing individual experiences as instrumental to the shared beliefs of multiculturalism, community, and the city the public concert audience becomes unified as "the group," allowing it, ultimately, to represent and reflect an ideal society, that of diverse Los Angeles. These moments of dancing audiences are described as important for Los Angeles, providing out-of-the-ordinary moments that are also fundamentally of and for the city. As a staff member said, "Grand Performances is about creating magical moments for the city of L.A.; it would be great if everyone in L.A. could experience this." Grand Performances' director's most memorable moment in the history of the series was the opening of the Water Stage in 1992. Poncho Sanchez, a Latin jazz artist, performed. The concert drew the largest crowd they had ever had; instead of the usual audience of two to three hundred people, three thousand people came.[4] The director described how L.A. was still feeling the impact of the 1992 riots and how he thought to himself at this event, "Isn't this the L.A. that we wish the rest of the world knew?"

Public concerts create a space for the recognition of an ideal that reflects the city at the same time as it is made possible by the city. Finding community at public concerts, or seeing the multicultural city in the diverse audience, creates public concerts as an ideal that has the very thing the city lacks. In the words of one board member, "In the end, downtown isn't about the buildings; it's a place for people to gather outdoors. It's free, fun, with people swaying to the music. At Grand Performances there's happiness, cohesion, and unity that's rare in a city of individualism." Grand Performances' director said that people had told him that Grand Performances is the humanization of L.A. and makes it possible to live in the city. Yet this lack—taking form as an absence of community, a sense of fragmentation, dehumanization, or a "city of individualism"—also produces a city reified in those terms. The aspect of ritual that makes it a "product of a more or less conflicted social reality" works here to "objectify conflict in the everyday world" (Comaroff 1985:119). Grand Performances' events stand outside everyday life, offering an experience that is counter to the norm. At public concerts, ideals of community are posited as counter to or solving problems of urbanism that make Los Angeles a city, creating it as having urban effects such as anomie and violence that require amelioration. The ideal must remain to some degree ephemeral, recognized only at events like public concerts, so that the urban can remain a problem in need of improvement. At the same time, it is necessary to believe that the transformative experiences accrued by attending public concerts, as utopic spaces, will extend to the city as a whole, and that society might indeed conform to the ideal created there.

The experience of dancing with strangers and recognizing difference that might be overcome is located in the context of the city. Something of the "universal" nature of dancing together, of embodied experience, facilitates a conceptual move from the audience to the city that glosses difference as diversity, celebrating a multicultural city that is experienced in an aspired-to moment of transcendence and togetherness. A longtime board member described a memorable event she had attended. Los Angeles is multicultural, she began, explaining that sixty languages are spoken at Belmont High School. Because downtown is neutral, she continued, people of different ethnic groups come to Grand Performances' concerts. Mike Davis, extolling the city's diversity, wrote that "eighty-six different languages were recently counted among its schoolchildren" (1992:81). Marking diversity as local and international, their statements posit a singularity of Los Angeles that is taken up by scholars, boosters, filmmakers, writers, artists, politicians, and residents who tout the city's social, architectural, and geographic diversity. At the same time, this piece of "multicultural evidence" ignores other things about the Los Angeles Unified School District, such as the fact that only 10 percent of the student body is white, though those who identified on their census forms as "non-Hispanic White" make up 30 percent of the population of the City of Los Angeles.[5] The absence of whites from Los Angeles public schools reflects a class divide that is mapped onto race and ethnicity, an effect of which is that white (and presumably wealthier) parents send their children to private schools at a higher percentage than does the rest of the population.

The concert, called "Cuatro Maestros," featured "Mexican-influenced music played in Texas bars from the 1940s, 1950s, 1960s." Recalling the evening, she described how there had been four different groups who began playing "early," around 5 or 6 in the evening, and did not finish until 10 or 11. By the time the last group played, many in the audience were dancing: mothers with their children and older couples, young parents and groups of women out for an evening of dancing. The music, she said, is "sort of like the equivalent of Tex-Mex food—it's a music that grew up there, and a lot of people dance to it." Passed down through generations even in the face of immigration, "a lot of Latinos from Mexico grew up with this music as children, from the age of two and three, dancing to this music with their parents and their grandparents and so on." During intermission she and her husband struck up a conversation with another couple, who told them, "Oh, yes, we grew up with this when we were little children." Food as a metaphor for music helps describe a cultural history that nonetheless immediately slips from something that could have a particular, geographically located history to one of authenticity in tradition. In narrating, she shifts from Tex-Mex to Mexico, from a locatable and recent development to a generalized "tradition" that goes back generations. This long, now Mexican, tradition of generations is reflected in Grand Performances' audience, expressed in particular by the couple next to the speaker and her husband. The move to Mexico from a description of music that was specifically Texan internationalizes an Angeleno multiculturalism at the same time as it provides a gloss of authenticity to those dancing. The geographic slippage, which might suggest either specific knowledge or a located understanding of Los Angeles, also projects "culture" as nationality.

When the music started again the couple invited her and her husband to dance, creating "a real social interaction across the various subcultures in L.A.," an "interaction among people that wouldn't have that interaction otherwise." Finishing, she said, "I think that's really important. Otherwise I don't think a city like L.A. will work. And I think that's the future, too, I think the world's going to be much more like that." Marked by skin color and ethnicity, persons with darker skin no longer only dance for whites as emblems of authentic culture as they did at the world's fairs of the late nineteenth and early twentieth centuries; instead, they help white people dance, enabling the experience through which the audience becomes a collective and a multicultural city is recognized. The Latino couple embodies a generational history that is now personal; the husband and wife "grew up with it" and taught it to the white couple, bringing them into that "tradition." Like other audience members, the board member extends her individual experience of dancing to the group, of a recognition of marked bodies to the terms of multiculturalism. With engagements across gender and ethnicity posited as being "across subcultures," public concerts become spaces in which fantasies of multicultural community are realized in discrete interactions with fellow audience members. Lasting memories of learning how to dance from people who learned from their parents and grandparents make real an ideal Los Angeles in which social diversity is experienced through dancing. Echoing the "future beats" of daKAH's MC, these aspirations of ideals of multicultural exchange and tolerance are projected onto the future of not only Los Angeles but the world.

Conclusion

Civic performance provides a space in which an understanding of music as something that brings people together is realized through embodied practices of listening, dancing, and gathering. Music spatializes experience, creating a resonant place that is outside time and everyday life. The recognition of the city in the collective of the audience becomes a means for those present to place themselves in Los Angeles, feeling a sense of civic belonging organized around people coming together from across the city to attend a downtown public concert. At public concerts, the performance is essential for what happens with the audience; the music has brought people together, has created the conditions whereby the audience will have diverse demographics, and gives people something to which to dance. Yet the audience turns to itself to ascribe meaning to the event, and the community "inspired" by the performances is recognized most obviously in the dancing audience.

The individual bodies, dancing and ethnically marked, are connected to and elided with the public body of the audience and of the city, creating a moral community of and for Los Angeles. The content of what is recognized is specific to a here and now of a neoliberal, global city and to a contemporary notion of multiculturalism that, in being recognized, helps produce those very things. In doing so it creates new kinds of urban subjects who, as members of a moral community that fosters tolerance over intolerance, consensus over dissent, and togetherness over fragmentation, enact and embody the comportment and disposition that facilitates urban civilities along those lines. As part of an experiential public of the performance, individuals express a feeling of having gained a sense of community with the group. An experience of dancing together at Grand Performances' public concerts reflects the work that has been done to foster harmony and consensus and the achievement of presenters' goals to create a space of multicultural civic unity. At the end of the night, after dancing with strangers and seeing people around them who signify diversity, some say that Grand Performances' audience is "Los Angeles at its best."

Notes

Introduction

1. Though the organization has had a series of name changes, I use "Grand Performances" throughout to avoid confusion, except when quoting archival documents that use an earlier name.

2. Grand Performances is not centrally concerned with presenting music that is specifically of Los Angeles. However, a wide-ranging scholarship on Los Angeles music addresses multifaceted relationships between race and ethnicity and musical expression (Bryant et al. 1998; DjeDje, Cogdell, and Meadows 1998; Loza 1993; Saul 2003). In considering questions of what is entailed in studying music in or of a city, how music contributes to city-making processes, and what kind of music emerges from urban social conditions, I draw especially from Allen and Wilcken 1998; Erlmann 1996; Finnegan 1989; Johnson 1995; C. Keil 1966; Nettl 1978; Shank 1994; Stokes 1992; and Strohm 1985.

3. Throughout the text I use a general term of "identity" or specify "ethnicity," "race," "nation," or "generation" depending on the term used in a given ethnographic context. I use "identity" loosely, insofar as it is necessary to have language to describe (though not account for) categories of ethnicity, race, and nationality that are deployed and remade at public concerts. I use "ethnic" and "national" in order to be both specific and broad. In specifying these terms I do not presume "ethnicity" and "nationality" to be coterminous.

4. The point here is not to provide a theorization of neoliberalism. I leave this task to others, drawing from recent work to understand some of its defining features (Brenner 2004; Brenner and Theodore 2002; Comaroff and Comaroff 2000; Harvey 2005). Relevant to the case of Grand Performances, it is generally agreed that neoliberalism has resulted in, among other things, the "elimination and/or intensified surveillance of urban public spaces" and the "creation of new privatized spaces of elite/ corporate consumption" (Brenner and Theodore 2002:24). Also pertinent are general trends of neoliberalism in cities that affect the nature of citizenship through processes such as gentrification, displacement, and segregation along lines of class and race, as well as increasing class disparities (Fainstein 2001; Holston 1989; Isin 2000; Leitner, Peck, and Sheppard 2006; Marcuse and Van Kempen 2000; M. Smith 1996).

5. Cities have long been recognized as important spaces for ongoing dynamics of globalization that take shape through macroeconomic and political processes imbricated in a range of local practices, institutional structures, and ideological iterations (Abrahamson 2004; Abu-Lughod 1999; Brenner 2004; Flusty 2003; Harvey 1990; R. Keil 1998; King 1997; Marcuse and Van Kempen 2000; Sassen 2000, 2001; M. Smith 2001). For discussions of Los Angeles as a global city, see Abu-Lughod 1999; Chinchilla and Hamilton 2001; Erie 2004; R. Keil 1998; and A. Scott 2000.

6. A history of diversity as a dominant defining feature of Los Angeles has yet to be written. Multiculturalism as a framework for describing the city's diversity emerged in the 1980s, marked most prominently by Mayor Tom Bradley's multiracial coalitions (Sonenshein 1993), a series of festivals that increasingly emphasized the performance of a diverse city, and artistic practice organized around a politics of identity and self (Cheng 2002; James 2003).

7. Los Angeles is defined broadly here, encompassing city, county, and the "urban agglomeration" that covers a variously defined swath of Southern California (L.A. Now 2001; Soja 1989). The specificity of "Los Angeles" is rarely elaborated at Grand Performances but is constituted through, among other things, audience-building and fund-raising; I take the liberty to use it in the same loose sense, elaborating the specificity of its making through the material as relevant.

8. Although free performances are discussed primarily in secondary accounts of the history of musical life in the United States (Starr 1987; Vaillant 2003), local municipal archives and other primary sources can often yield information about motivations for and debates around free performances from the last century in the context of other discussions.

9. Using "civic subject" as opposed to "informal citizenship" (Sassen 2003), "flexible citizenship" (Ong 1999), or "cultural citizenship" (Joseph 1999; Ong 1996) allows for the inclusion and study of a range of practices as significant for the formation of urban subjects, without needing to address these subjects' formal citizenship status.

10. This project draws on literature on the intersections of music, space, and identity (Bull and Back 2003; Connell and Gibson 2003; Krims 2007; Leyshon, Matless, and Revill 1998; Krims 2007; Stokes 1994; Whiteley, Bennett, and Hawkins 2004); rather than emphasizing discrete ethnic or racial groups and their musical forms, it explores how these themes play out on the level of the city as a whole, and how discrete groups are drawn on for a multicultural project. A wealth of work in ethnomusicology provides tools for situating this project in broader frameworks of the social practices and meanings of music. Scholarship that has informed my analysis includes Blacking 1995; Feld 1990; Fox 2004; Seeger 2004; Small 1998; and Stokes 1992.

11. As Bourdieu (1993) posits, the "field of cultural production" has its own logic and is not epiphenomenal as ideology (Kruger 1992:8–9). Following Foucault's "rule of double conditioning" (1978:99–100), an institutional form can help produce wider social processes without being subsumed by those processes.

12. According to Bakhtin, a utopic experience can be seen in medieval carnival; there, "Free, familiar contacts were deeply felt and formed an essential element of the carnival spirit. People were, so to speak, reborn for new, purely human relations. These truly human relations were not only a fruit of imagination or abstract thought; they were experienced. The utopian ideal and the realistic merged in this carnival experience, unique in its kind" (1984:10).

Chapter 1

1. Davis stresses the geographic class divide of downtown Los Angeles that the Bunker Hill Urban Renewal Project both responded to and helped create, explaining how "At immense public cost, the corporate headquarters and financial district was shifted from the old Broadway-Spring corridor six blocks west to the Greenfield site created by destroying the Bunker Hill residential neighborhood" (1992:230). The resulting effect, he says, was "spatial apartheid" with Hill Street providing a "local Berlin wall" between California Plaza and the "lifeworld of Broadway, now reclaimed by Latino immigrants as their primary shopping and entertainment street" (1992:230).

2. The protests of historic preservationists motivated the CRA/LA to move some of the mansions, which are now in the Heritage Square Museum at Avenue 43 just off the Pasadena freeway, five miles northeast of downtown Los Angeles.

3. The Community Redevelopment Law that outlines the formation and nature of redevelopment agencies is found in Division 24, Part I (Section 33000 et seq.) of the California Health and Safety Code. According to Mel Scott, "In 1945 the California Legislature enacted a Community Redevelopment Act authorizing creation of local redevelopment agencies to acquire blighted properties (through condemnation if necessary), clear them, and resell or lease them to private enterprise to rebuild in accordance with the master plan of the community and in conformity with standards established by the local planning commission" (1949:110). Accused at times of being a semi-private organization with no public oversight, today the CRA/LA is managed by a board of seven commissioners appointed by the mayor; its work is overseen by the City Council (CRA/LA n.d.:2). At the time of this writing, the manager of the CRA/LA's Bunker Hill project was also on Grand Performances' board.

4. Bunker Hill Associates was a joint venture of Cadillac-Fairview/California, Goldrich, Kest & Associates, and Shapell Government Housing, Inc. According to people involved at the time, the team was chosen over the young Los Angeles–based development team of Maguire-Thomas primarily because it was able to show it had the funds to complete the project.

5. Arco, or the Atlantic Richfield Company, had its headquarters on Bunker Hill for many years and was a major arts supporter. After BP (British Petroleum) purchased Arco, the company moved out of downtown Los Angeles. Arco's eponymous building was subsequently renamed after City National Bank ("Arco Plaza to Be Renamed" 2004).

6. Free concerts in California Plaza were initially supported and run by Metropol-

itan Structures West, the original owner and manager of One California Plaza, the first office tower to be completed. After the second office tower was built an independent organization was formed to manage the common area and produce the performances. The programming director shifted to a position as head of Plaza Commons, Inc. (PCI), the management company for the public area of the corporate plaza. The CRA/LA required that a set amount of PCI's budget be used for the concerts. In 1990 the city provided support for the evening series, while the building paid for thirty-five lunch-hour concerts.

7. In accordance with the CRA/LA requirement that the building owners support the concerts, funds continue to be drawn from parking garage revenue to provide roughly half of Grand Performances' million-dollar annual budget.

8. PCI provided start-up costs and services for Friends of California Plaza Presents. In 1997 the building owners went to the CRA/LA and requested that their required contribution to the concerts through PCI be reduced. Reflecting the economic conditions at the time, the office towers were not fully leased, and the building owners felt their contribution should be scaled in relation to their income. Moreover, income from filming in the Plaza had exceeded expectations that year. They were allowed to reduce their contribution by $189,000. Hence annual support of $900,000 was reduced to $739,000, $50,000 of which is used to support Angel's Flight, a reproduction of the original funicular railroad that carried people up Bunker Hill. That year the CRA/LA gave a one-time grant to Friends of California Plaza Presents of $150,000, with the expectation that after that point the concert presenters would be able to raise that amount of money. In 1998 the season was reduced to eighteen evening concerts when the CRA/LA grant was not renewed. In 2003, with the office towers fully rented, the concert presenters requested an increase in support, such that their budget would be restored to its earlier level. Even after pressure from the CRA/LA project manager, the PCI board refused.

9. The name of the organization at the time was California Plaza Presents and the nonprofit was Friends of California Plaza Presents. The original minutes are in Grand Performances' office and copies are in the author's files.

Chapter 2

1. The use of terms and their capitalization follows that of the book.

2. Following a trajectory similar to that of KSCI, KIRN was started in 1999 by Howard Kalmenson, who, noticing the area marked "Little Tehran" on the Westside, thought it would be a lucrative endeavor to start a radio station that catered to Iranian immigrants. Kalmenson owns thirty radio stations, including two Spanish-language stations he bought in the 1960s "because he sensed a growing market in Los Angeles" (Mena 2002:B14).

3. The Music Performance Fund (MPF) is a compelling example of an exchange of consumer capital for public concert funding. Created in 1948 as part of a labor agreement between the American Federation of Musicians, the national musicians'

union, and the Recording Industry Association of America to address the loss of jobs for musicians when recordings replaced live musicians on the radio, the MPF uses a percentage of all sales of recorded music to support free concerts across the United States and Canada.

Chapter 3

1. Musical genre is constituted through intertextual processes that structure how a text is interpreted as part of a particular genre and as indexical of other social meanings in a range of contexts (Bauman 2004; Briggs and Bauman 1992; Hutcheon 2000; Kapchan 1995). For musical genre, intertextuality includes institutional framings (Negus 1999), negotiations in performance (Holt 2007), pragmatic and metapragmatic indexical markers (Keil and Feld 1994; Monson 1996), historical conditions, and demographics of performers and audience. These are unevenly drawn on to frame any particular performance. Genre-formation is an intersubjective process, such that a relationship between performer and audience authorizes daKAH as hip-hop and as orchestra.

2. For some time daKAH was sponsored by the car company Scion, which markets its vehicles with a hip-hop image that is presented through commercials featuring hip-hop artists, compilation CDs of various DJs and hip-hop groups, and gallery shows for graffiti artists. Its sponsorship of daKAH has included funding a studio recording and paying for the orchestra to travel to Austin, Texas, to play at the annual music festival South by Southwest.

3. A multiscalar production of locality (Appadurai 1996) is a central element of hip-hop lyrics (Forman 2002; Rose 1994) as well as many other kinds of music (Cohen and Fairley 2000; Leyshon, Matless, and Revill 1998; Stokes 1994).

Chapter 4

1. As Brown expands, "A formulation of the political state and of citizenship that . . . abstracts from the substantive conditions of our lives, works to prevent recognition or articulation of differences *as* political—as effects of power—in their very construction and organization . . . the universality of the state is ideologically achieved by turning away from and thus depoliticizing, yet at the same time *presupposing*, our collective particulars, not by embracing them, let alone emancipating us from them. In short, 'the political' in liberalism is precisely not a domain for social identification: expected to recognize our political selves in the state, we are not led to expect deep recognition there" (1996:153–54).

2. The "uniqueness" of music is inseparable from its interpretation and wider social meanings. As Deutsche suggests, "No order of meaning exists in itself. No founding presence, unconditional source, or privileged ground guarantees the authority of meanings constituted apart from discursive interventions" (1996:230). While many have explored the question of how music—as a special, sensory, and participatory medium—can represent intra- and extra-musical issues or be represented in non-

musical terms (Keil and Feld 1994; Monson 1996; Stokes 1994; Turino 1999), unpacking musical meaning remains a "tantalizing, confusing, problematic area of inquiry" (Maus 2001:642).

3. I am grateful to Adel Wang for bringing this to my attention.

4. In 2003 Culture Clash presented a play about the history of Chavez Ravine at the Mark Taper Forum that relayed the area's transformation from poor community through urban renewal and the building of Dodger Stadium. The performance and concurrent public events reflected ways in which the group combines art and politics. A panel discussion was attended by affordable housing advocates who used the forum to pass out flyers and lobby other audience members to participate in their work.

Chapter 5

1. According to Warner, publics, constituted through the circulation of texts, require moments in which they come together in order to see themselves. Always partial, those moments are nonetheless key for validating experiences of belonging and membership. Though Warner (2002) presents these events as secondary to the socialities produced through circulation, the articulation of sentiments of civic belonging and aspirations in the context of public concerts suggests that audiences of live performances should be taken seriously in considering how and where meaning making occurs within the formation of publics.

2. "Scholars, ritual inventors, and ritual participants do not usually see how scholarship has constructed this notion of ritual or the type of authority it has acquired. They think of 'ritual in general' as something that has been there all along and only now discovered—no matter whether it is thought to be a social constant, a psychological necessity, or a biological determinism. As an abstraction that determines all particular rites and ceremonies, ritual itself becomes a reified construct with the authority to sanction new forms of ritualization that appeal to it as a quintessential human and social dynamic" (Bell 1997:263).

3. Beeman (1993:384) distinguishes between audience as participant, as witness, and as evaluator, claiming that in the Western concert setting, the audience is only witness and evaluator, whereas in a non-Western ritual setting the audience is also a participant, giving things to deities, interacting with priests, and so forth. This schema presumes a center-periphery divide in which the "periphery" has culture and participation through visible activity, and the "center" does not. Yet audience participation—whether as silent "witness," dancing, or walking around the perimeter of the Plaza—is integral to the public concert performance, which would not happen without it.

4. Grand Performances' ability to realize ideals of a diverse city through performance is achieved through the iteration and articulation of the recognition of those ideals by some present rather than a quantifiable relationship to the city's population.

5. "Non-Hispanic White" is a category in the U.S. Census that reflects a definition of Hispanic as an "ethnicity" rather than a "race" or "nationality"; because those identifying as Hispanic can choose another racial category, "non-Hispanic White" is used to mark a person who is White with no Hispanic background.

Bibliography

Abrahamson, Mark. 2004. *Global Cities*. Oxford: Oxford University Press.

Abu-Lughod, Janet L. 1999. *New York, Chicago, Los Angeles: America's Global Cities*. Minneapolis: University of Minnesota Press.

Acocella, Joan, Lynn Garafola, and Jonnie Greene. 1987. The Rite of Spring Considered as a Nineteenth-Century Ballet. In *The Rite of Spring at Seventy-Five*, 3–5. New York: New York Council for the Humanities.

Adler, Patricia. 1964. *The Bunker Hill Story*. Glendale, Calif.: La Siesta Press.

Adorno, Theodor. 1994. *Philosophy of Modern Music*. Trans. Anne G. Mitchell and Wesley V. Blomster. New York: Continuum.

Allen, James P., and Eugene Turner. 1997. *The Ethnic Quilt: Population Diversity in Southern California*. Northridge: Center for Geographical Studies, California State University.

———. 2002. *Changing Faces, Changing Places: Mapping Southern California*. Northridge: Center for Geographical Studies, California State University.

Allen, Ray, and Lois Wilcken, eds. 1998. *Island Sounds in the Global City: Caribbean Popular Music and Identity in New York*. Brooklyn, N.Y.: Institute for Studies in American Music, Brooklyn College.

Appadurai, Arjun. 1996. *Modernity at Large: Cultural Dimension of Globalization*. Minneapolis: University of Minnesota Press.

Arco Plaza to Be Renamed. 2004. *Los Angeles Downtown News*, September 20:2.

Arthur Erickson Architects. 1985. *California Plaza Phase 2A Program Central Plaza*. Los Angeles: Arthur Erickson Architects, WZMH Group.

Asian Pacific American Legal Center of Southern California. 2008. *L.A. Speaks: Language Diversity and English Proficiency by Los Angeles County Service Planning Area*. Los Angeles: Asian Pacific American Legal Center.

Attali, Jacques. 1999. *Noise: The Political Economy of Music*. Trans. Brian Massumi. Minneapolis: University of Minnesota Press.

Bakhtin, Mikhail. 1984. *Rabelais and His World*. Bloomington: Indiana University Press.

Banfield, Edward C. 1984. *The Democratic Muse*. New York: Basic Books.

Barber, Karen. 1997. Preliminary Notes on Audiences in Africa. *Africa* 67:347–62.

Barthes, Roland. 1972. *Mythologies*. Trans. Annette Lavers. New York: Hill and Wang.

———. 1985. *The Responsibility of Forms.* Trans. Richard Howard. Berkeley: University of California Press.

Bauman, Richard. 2004. *A World of Others' Words: Cross-Cultural Perspectives on Intertextuality.* Oxford: Blackwell.

Beeman, William O. 1993. The Anthropology of Theater and Spectacle. *Annual Review of Anthropology* 22:369–93.

Bell, Catherine. 1992. *Ritual Theory, Ritual Practice.* New York: Oxford University Press.

———. 1997. *Ritual: Perspectives and Dimensions.* Oxford: Oxford University Press.

Benedict, Stephen, ed. 1991. *Public Money & the Muse: Essays on Government Funding for the Arts.* New York: Norton.

Berelowitz, Jo-Anne. 1991. LA Stories: Of Art, MOCA, and City-Building. Ph.D. thesis, University of California–Los Angeles.

Bianchini, Franco. 1990. Urban Renaissance? The Arts and the Urban Regeneration Process. In *Tackling the Inner Cities: The 1980s Reviewed, Prospects for the 1990s,* ed. Susanne MacGregor and Ben Pimlott, 215–50. Oxford: Clarendon Press.

Bijsterveld, Karin. 2008. *Mechanical Sound: Technology, Culture, and Public Problems of Noise in the Twentieth Century.* Cambridge, Mass.: MIT Press.

Blacking, John. 1995. *How Musical Is Man?* Seattle: University of Washington Press.

Bonacich, Edna, Paul M. Ong, and Lucie Cheng, eds. 1994. *The New Asian Immigration in Los Angeles and Global Restructuring.* Philadelphia: Temple University Press.

Born, Georgina, and David Hesmondhalgh, eds. 2000. *Western Music and Its Others: Difference, Representation, and Appropriation in Music.* Berkeley: University of California Press.

Bourdieu, Pierre. 1984. *Distinction: A Social Critique of the Judgement of Taste.* Trans. Richard Nice. Cambridge, Mass.: Harvard University Press.

———. 1993. *The Field of Cultural Production: Essays on Art and Literature.* New York: Columbia University Press.

Boyer, Christine M. 1983. *Dreaming the Rational City: The Myth of American City Planning.* Cambridge, Mass.: MIT Press.

Brenner, Neil. 2004. *New State Spaces: Urban Governance and the Rescaling of Statehood.* Oxford: Oxford University Press.

Brenner, Neil, and Nik Theodore. 2002. Cities and the Geographies of "Actually Existing Neoliberalism." In *Spaces of Neoliberalism: Urban Restructuring in North America and Western Europe,* ed. Neil Brenner and Nik Theodore, 2–32. Oxford: Blackwell.

Briggs, Charles, and Richard Bauman. 1992. Genre, Intertextuality, and Social Power. *Journal of Linguistic Anthropology* 2:131–72.

Brown, Wendy. 1996. Injury, Identity, Politics. In *Mapping Multiculturalism,* ed. Avery F. Gordon and Christopher Newfield, 149–66. Minneapolis: University of Minnesota Press.

Brubaker, Rogers, and Frederick Cooper. 2000. Beyond "Identity." *Theory and Society* 29:1–47.

Bryant, Clora, Buddy Collette, William Green, Steven Isoardi, Jack Kelson, Horace Tapscott, Gerald Wilson, and Marl Young, eds. 1998. *Central Avenue Sounds: Jazz in Los Angeles.* Berkeley: University of California Press.

Bull, Michael, and Les Back, eds. 2003. *The Auditory Culture Reader.* Oxford: Berg.

Bunker Hill Associates. 1980. *A Proposal for Bunker Hill.* Los Angeles: Bunker Hill Associates.

———. 1982. *Plaza for the Performing Arts: Phase 2, Preliminary Outline Operating Plan, California Plaza, Performing Arts Plaza.* Los Angeles: Bunker Hill Associates.

———. 1987. *California Plaza: The Arts in an Urban Garden.* Los Angeles: Bunker Hill Associates.

Bunker Hill Task Force. 1980. Minority Report. Bunker Hill Collection. Doheny Memorial Library, University of Southern California.

Butler, Judith. 1993. *Bodies That Matter: On the Discursive Limits of "Sex."* New York: Routledge.

Cage, John. 1973. *Silence: Lectures and Writings.* Middletown, Conn.: Wesleyan University Press.

Campbell, Gavin James. 2000. A Higher Mission Than Merely to Please the Ear: Music and Social Reform in America, 1900–1925. *Musical Quarterly* 84:259–86.

Carney, Steve. 2003. Hour by Hour in the Ratings War. *Los Angeles Times*, April 25:E29.

Castles, Stephen, and Alastair Davidson. 2000. *Citizenship and Migration: Globalization and the Politics of Belonging.* New York: Routledge.

Cheng, Meiling. 2002. *In Other Los Angeleses: Multicentric Performance Art.* Berkeley: University of California Press.

Chinchilla, Norma, and Nora Hamilton. 2001. *Seeking Community in the Global City: Salvadorans and Guatemalans in Los Angeles.* Philadelphia: Temple University Press.

Clifford, James, and George E. Marcus, eds. 1986. *Writing Culture.* Berkeley: University of California Press.

Cohen, Sara, and Jan Fairley, eds. 2000. Place issue of *Popular Music* 19(1).

Comaroff, Jean. 1985. *Body of Power, Spirit of Resistance: The Culture and History of a South African People.* Chicago: University of Chicago Press.

Comaroff, Jean, and John L. Comaroff. N.d. Reflections on Youth: From the Past to the Postcolony. Unpublished manuscript.

———. 2000. Millennial Capitalism: First Thoughts on a Second Coming. *Public Culture* 12(2):291–343.

Comaroff, John, and Jean Comaroff. 1992. *Ethnography and the Historical Imagination.* Boulder, Colo.: Westview Press.

Community Redevelopment Agency of the City of Los Angeles, California (CRA/LA).

N.d. *Partners in Revitalizing Communities.* Los Angeles: Community Redevelopment Agency.

———. 1958. *Redevelopment Plan for the Bunker Hill Urban Renewal Project 1B a Part of Central Redevelopment Area 1.* Los Angeles: Community Redevelopment Agency.

———. 1959. *Amended Redevelopment Plan, Bunker Hill Urban Renewal Project (California R-1).* Los Angeles: Community Redevelopment Agency.

———. 1979. *Development Offering, Remaining 8.75 Acres of Bunker Hill, Downtown Los Angeles.* Los Angeles: Community Redevelopment Agency.

———. 1981. *Bunker Hill Urban Renewal Project: Disposition and Development Agreement, RSTUY-1.* Los Angeles: Community Redevelopment Agency.

———. 1990. *Bunker Hill Redevelopment Project Biennial Report, 1988–1990.* Los Angeles: Community Redevelopment Agency.

———. 2000. *Bunker Hill Urban Renewal Project Proposed Five-Year Implementation Plan FY2000–FY2004.* Los Angeles: Community Redevelopment Agency.

———. 2005. Frequently Asked Questions. http://www.crala.org/internet-site/faqs.cfm (accessed April 26, 2008).

Condry, Ian. 2006. *Hip-Hop Japan: Rap and the Paths of Cultural Globalization.* Durham, N.C.: Duke University Press.

Connell, John, and Chris Gibson. 2003. *Sound Tracks: Popular Music, Identity and Place.* London: Routledge.

Connolly, Julie, Michael Leach, and Lucas Walsh, eds. 2007. *Recognition in Politics: Theory, Policy and Practice.* Newcastle, U.K.: Cambridge Scholars Publishing.

Crawford, Dorothy L. 1995. *Evenings on and off the Roof: Pioneering Concerts in Los Angeles, 1939–1971.* Berkeley: University of California Press.

Crouch, Dora Polk. 1981. The Historical Development of Urban Open Space. In *Urban Open Spaces,* ed. Lisa Taylor, 7–8. New York: Rizzoli.

D'Accone, Frank A. 1997. *The Civic Muse: Music and Musicians in Siena During the Middle Ages and the Renaissance.* Chicago: University of Chicago Press.

Dávila, Arlene. 2001. *Latinos, Inc.: The Marketing and Making of a People.* Berkeley: University of California Press.

Davis, Mike. 1992. *City of Quartz: Excavating the Future in Los Angeles.* New York: Vintage Books.

Dear, Michael. 1986. Postmodernism and Planning. *Society & Space: Environment & Planning D* 4:367–84.

———. 2000. *The Postmodern Urban Condition.* Oxford: Blackwell.

———, ed. 2002. *From Chicago to L.A.: Making Sense of Urban Theory.* London: Sage.

Dear, Michael J., H. Eric Schockman, and Greg Hise, eds. 1996. *Rethinking Los Angeles.* Thousand Oaks, Calif.: Sage.

de Certeau, Michel. 2002. *The Practice of Everyday Life.* Berkeley: University of California Press.

Der-martirosian, Claudia. 2007. *Iranian Immigrants in Los Angeles: The Role of Networks and Economic Integration.* New York: LFB Scholarly Publishing.

Derrida, Jacques. 1994. *Specters of Marx.* New York: Routledge.

Deutsche, Rosalyn. 1996. *Evictions: Art and Spatial Politics.* Cambridge, Mass.: MIT Press.

Dickinson, Edward. 1919. Ideals in Music. In *Ideals of America: Analyses of the Guiding Motives of Contemporary American Life by Leaders in Various Fields of Thought and Action.* Prepared for the City Club of Chicago, 1916–19. Chicago: A. C. McClurb and Company.

DjeDje, Jacqueline Cogdell, and Eddie S. Meadows, eds. 1998. *California Soul: Music of African Americans in the West.* Berkeley: University of California Press.

Durkheim, Emile. 1915. *The Elementary Forms of Religious Life.* Trans. Joseph Ward Swain. New York: Free Press.

Erie, Steven P. 2004. *Globalizing L.A.: Trade, Infrastructure, and Regional Development.* Stanford, Calif.: Stanford University Press.

Erlmann, Veit. 1996. *Nightsong: Performance, Power, and Practice in South Africa.* Chicago: University of Chicago Press.

Estrada, William David. 2008. *The Los Angeles Plaza: Sacred and Contested Space.* Austin: University of Texas Press.

Fabian, Johannes. 1983. *Time and the Other: How Anthropology Makes Its Object.* New York: Columbia University Press.

Fainstein, Susan. 2001. *City Builders: Property Development in New York and London, 1980–2000.* Lawrence: University Press of Kansas.

Feld, Steven. 1990. *Sound and Sentiment: Birds, Weeping, Poetics, and Song in Kaluli Expression.* Philadelphia: University of Pennsylvania Press.

———. 1994. Communication, Music, and Speech About Music. In *Music Grooves*, ed. Charles Keil and Steven Feld, 77–95. Chicago: University of Chicago Press.

———. 2000. Sound Worlds. In *Sound*, ed. Patricia Kruth and Henry Stobart, 173–200. Cambridge: Cambridge University Press.

Finnane, Antonia. 2008. *Changing Clothes in China: Fashion, History, Nation.* New York: Columbia University Press.

Finnegan, Ruth. 1989. *The Hidden Musicians.* Cambridge: Cambridge University Press.

Flores, Juan. 2004. Puerto Rocks: Rap, Roots, and Amnesia. In *That's the Joint! The Hip Hop Studies Reader*, ed. Murray Forman and Mark Anthony Neal, 69–86. New York: Routledge.

Flusty, Steven. 2003. *De-Coca-Colonization: Making the Globe from the Inside Out.* New York: Routledge.

Fogelson, Robert M. 2001. *Downtown: Its Rise and Fall, 1880–1950.* New Haven, Conn.: Yale University Press.

Forman, Murray. 2002. *The 'Hood Comes First: Race, Space, and Place in Rap and Hip-Hop.* Middletown, Conn.: Wesleyan University Press.

Forman, Murray, and Mark Anthony Neal, eds. 2004. *That's the Joint! The Hip Hop Studies Reader.* New York: Routledge.

Foucault, Michel. 1977. *Discipline and Punish: The Birth of the Prison.* Trans. Alan Sheridan. New York: Vintage Books.

————. 1978. *History of Sexuality, Volume 1: An Introduction.* New York: Vintage Books.

Fox, Aaron. 2004. *Real Country: Music and Language in Working-Class Culture.* Durham, N.C.: Duke University Press.

Fraser, Nancy, and Axel Honneth. 2003. *Redistribution or Recognition? A Political-Philosophical Exchange.* London: Verso.

Frith, Simon. 1996. Music and Identity. In *Questions of Cultural Identity*, ed. Stuart Hall and Paul du Gay, 108–27. London: Sage.

Gang Starr. 1991. Liner notes. *Step in the Arena.* Chrysalis 21798.

Garden Theatre Festival. Papers. Charles E. Young Research Library, University of California–Los Angeles.

Ginsburg, Faye D., Lila Abu-Lughod, and Brian Larkin, eds. 2002. *Media Worlds: Anthropology on New Terrain.* Berkeley: University of California Press.

Gladney, Dru C. 1994. Representing Nationality in China: Refiguring Majority/Minority Identities. *Journal of Asian Studies* 53(1):92–123.

Gold, Matea. 2003. Citing Stigma, L.A. May Drop Name "South-Central." *Los Angeles Times*, April 9:A1, A24.

Gold, Matea, and Greg Braxton. 2003. Considering South-Central by Another Name. *Los Angeles Times*, April 10:B3.

Goldstein, Barbara. 1980. News Report. *Progressive Architecture* (September:46).

Gordon, Colin. 2004. Blighting the Way: Urban Renewal, Economic Development, and the Elusive Definition of Blight. *Fordham Urban Law Journal* 31:305–37.

Greenberg, Steve. 1999. Sugar Hill Records. In *The Vibe History of Hip Hop*, ed. Alan Light, 23–33. New York: Three Rivers Press.

Gupta, Akhil, and James Ferguson, eds. 1997. *Anthropological Locations: Boundaries and Grounds of a Field Science.* Berkeley: University of California Press.

Habell-Pallán, Michelle. 1999. El Vez Is "Taking Care of Business": The Inter/National Appeal of Chicano Popular Music. *Cultural Studies* 13(2):195–210.

Hall, Peter Dobkin. 1992. *Inventing the Nonprofit Sector and Other Essays on Philanthropy, Voluntarism, and Nonprofit Organizations.* Baltimore: Johns Hopkins University Press.

Hammack, David C., ed. 1998. *Making the Nonprofit Sector in the United States: A Reader.* Bloomington: Indiana University Press.

Hapeta, Dean aka Te Kupu. 2003. *Ngâtahi: Know the Links.* Kia Kaha Productions.

Harvey, David. 1990. *The Condition of Postmodernity: An Enquiry into the Origins of Cultural Change.* Cambridge, Mass.: Blackwell.

————. 2001. *Spaces of Capital: Towards a Critical Geography.* London: Routledge.

————. 2005. *A Brief History of Neoliberalism.* Oxford: Oxford University Press.

Hebdige, Dick. 1981. *Subculture: The Meaning of Style.* New York: Routledge.

Hegel, G. W. F. 1991. *Elements of the Philosophy of Right.* Ed. Allen W. Wood. Trans. H. B. Nisbet. Cambridge: Cambridge University Press.

Hoeckner, Berthold. 2002. *Programming the Absolute: Nineteenth-Century German Music and the Hermeneutics of the Moment.* Princeton, N.J.: Princeton University Press.

Hollinger, David A. 2000. *Postethnic America: Beyond Multiculturalism.* Rev. ed. New York: Basic Books.

Holston, James. 1989. *The Modernist City: An Anthropological Critique of Brasília.* Chicago: University of Chicago Press.

Holston, James, and Arjun Appadurai. 1996. Cities and Citizenship. *Public Culture* 8:187–204.

Holt, Fabian. 2007. *Genre in Popular Music.* Chicago: University of Chicago Press.

Hutcheon, Linda. 2000. *A Theory of Parody: The Teachings of Twentieth-Century Art Forms.* Urbana: University of Illinois Press.

Huyssen, Andreas. 1986. *After the Great Divide: Modernism, Mass Culture, Postmodernism.* Bloomington: University of Indiana Press.

Hylen, Arnold. 1976. *Bunker Hill: A Los Angeles Landmark.* Los Angeles: Dawson's Book Shop.

Isin, Engin F. 1992. *Cities Without Citizens: The Modernity of the City as a Corporation.* Montreal: Black Rose Books.

———, ed. 2000. *Democracy, Citizenship and the Global City.* London: Routledge.

Isin, Engin F., and Greg M. Nielsen, eds. 2008. *Acts of Citizenship.* London: Zed Books.

Isoardi, Steven L. 2006. *The Dark Tree: Jazz and the Community Arts in Los Angeles.* Berkeley: University of California Press.

James, David E., ed. 2003. *The Sons and Daughters of Los: Culture and Community in L.A.* Philadelphia: Temple University Press.

Jameson, Fredric. 1991. *Postmodernism, or, The Cultural Logic of Late Capitalism.* Durham, N.C.: Duke University Press.

Jencks, Christopher. 1993. *Heterotopolis: Los Angeles, the Riots and the Strange Beauty of Hetero-Architecture.* London: Academy Editions.

Johnson, James H. 1995. *Listening in Paris: A Cultural History.* Berkeley: University of California Press.

Jonas, Andrew E. G., and David Wilson, eds. 1999. *The Urban Growth Machine: Critical Perspectives, Two Decades Later.* Albany: State University of New York Press.

Joseph, May. 1999. *Nomadic Identities: The Performance of Citizenship.* Minneapolis: University of Minnesota Press.

Kapchan, Deborah. 1995. Performance. *Journal of American Folklore* 108(430):478–508.

———. 2008. The Promise of Sonic Transformation: Performing the Festive Sacred in Morocco. *American Anthropologist* 110(4):467–83.

Kaplan, David E. 1992. *Fires of the Dragon: Politics, Murder, and the Kuomintang.* New York: Atheneum.

Katz, Mark. 2004. *Capturing Sound: How Technology Has Changed Music.* Berkeley: University of California Press.

KAZN AM1300. n.d. About KAZN AM1300. http://www.am1300.com/ (accessed July 25, 2009).

Keil, Charles. 1966. *Urban Blues.* Chicago: University of Chicago Press.

Keil, Charles, and Steven Feld. 1994. *Music Grooves.* Chicago: University of Chicago Press.

Keil, Roger. 1998. *Los Angeles: Globalization, Urbanization and Social Struggles.* New York: John Wiley and Sons.

Keith, Michael, and Steve Pile, eds. 1993. *Place and the Politics of Identity.* London: Routledge.

Kelley, Robin D. G. 1994. *Race Rebels: Culture, Politics, and the Black Working Class.* New York: The Free Press.

Kelly, Thomas Forrest. 2000. *First Nights: Five Musical Premieres.* New Haven, Conn.: Yale University Press.

King, Anthony D., ed. 1997. *Culture, Globalization and the World-System: Contemporary Conditions for the Representation of Identity.* Minneapolis: University of Minnesota Press.

King, Russell, and Nancy Wood, eds. 2001. *Media and Migration: Constructions of Mobility and Difference.* New York: Routledge.

Kirshenblatt-Gimblett, Barbara. 1995. Confusing Pleasures. In *The Traffic in Culture: Refiguring Art and Anthropology*, ed. George E. Marcus and Fred R. Myers, 224–55. Berkeley: University of California Press.

KMRB AM1430. 2003. Background of KMRB AM1430. http://www.am1430.net/ (accessed July 25, 2009).

Kohpahl, Gabriel. 1998. *Voices of Guatemalan Women in Los Angeles: Understanding Their Immigration.* New York: Routledge.

Kotler, Philip. 2002. *Marketing Places.* New York: Free Press.

Kreiswirth, Sandra. 1993. Home Grown. *Torrance Daily Breeze*, August 13:K14, K16, K24.

Krims, Adam. 2000. *Rap Music and the Poetics of Identity.* Cambridge: Cambridge University Press.

———. 2007. *Music and Urban Geography.* New York: Routledge.

Kruger, Loren. 1992. *The National Stage: Theatre and Cultural Legitimation in England, France, and America.* Chicago: University of Chicago Press.

KSCI-TV. 2002. *KSCI Research Brief. 2nd Quarter.* KSCI-TV: Los Angeles.

———. 2006. About Us—Coverage Map. La18tv. http://www.la18.tv/aboutus_coverage map.aspx (accessed March 9, 2009).

KSCI-TV Reaches Millions of Asian-Americans. 2002. *Advertising Age*, November 4.

Kymlicka, Will. 1995. *Multicultural Citizenship.* Oxford: Oxford University Press.

L.A. Now. 2001. *L.A. Now*. Vol. 1. Pasadena, Calif.: Art Center College of Design.

Larkin, Brian. 2008. *Signal and Noise*. Durham, N.C.: Duke University Press.

Lefebvre, Henri. 1991. *The Production of Space*. Trans. Donald Nicholson-Smith. Oxford: Basil Blackwell.

———. 1996. *Writings on Cities*. Trans. and ed. Eleonore Kofman and Elizabeth Lebas. Oxford: Blackwell.

———. 2004. *Rhythmanalysis*. New York: Continuum.

Leitner, Helga, Jamie Peck, and Eric Sheppard, eds. 2006. *Contesting Neoliberalism: Urban Frontiers*. New York: Guilford Press.

Leyshon, Andrew, David Matless, and George Revill. 1998. *The Place of Music*. New York: Guilford Press.

Light, Alan. 2004. About a Salary or Reality? Rap's Recurrent Conflict. In *That's the Joint! The Hip Hop Studies Reader*, ed. Murray Forman and Mark Anthony Neal, 137–45. New York: Routledge.

Logan, John, and Harvey Molotch. 1987. *Urban Fortunes: The Political Economy of Place*. Berkeley: University of California Press.

Los Angeles Police Department. 2004–8. Noise Enforcement Team. http://www.lapdonline.org/special_operations_support_division/content_basic_view/1031 (accessed February 20, 2009).

Loukaitou-Sideris, Anastasia and Tridib Banerjee. 1998. *Urban Design Downtown: Poetics and Politics of Form*. Berkeley: University of California Press.

Loukaitou-Sideris, Anastasia, and Gail Sansbury. 1996. Lost Streets of Bunker Hill. *California History* 74(4):394–407, 448–49.

Low, Setha, Dana Taplin, and Suzanne Scheld. 2005. *Rethinking Urban Parks: Public Space and Cultural Diversity*. Austin: University of Texas Press.

Lowe, Lisa. 1996. Imagining Los Angeles in the Production of Multiculturalism. In *Mapping Multiculturalism*, ed. Avery Gordon and Christopher Newfield, 413–23. Minneapolis: University of Minnesota Press.

Loza, Steven. 1993. *Barrio Rhythms: Mexican American Music in Los Angeles*. Urbana: University of Illinois Press.

Lubiano, Wahneema. 1996. Like Being Mugged by a Metaphor: Multiculturalism and State Narrative. In *Mapping Multiculturalism*, ed. Avery F. Gordon and Christopher Newfield, 64–75. Minneapolis: University of Minnesota Press.

Marcus, George. 2008. The End(s) of Ethnography: Social/Cultural Anthropology's Signature Form of Producing Knowledge in Transition. *Cultural Anthropology* 3(1):1–14.

Marcuse, Peter, and Ronald Van Kempen. 2000. *Globalizing Cities: A New Spatial Order*. Oxford: Blackwell.

Markell, Patchen. 2003. *Bound by Recognition*. Princeton, N.J.: Princeton University Press.

Martin, Randy. 2004. Dance and Its Others: Theory, State, Nation, and Socialism. In

Of the Presence of the Body: Essays on Dance and Performance Theory, ed. André Lepecki, 47–63. Middletown, Conn.: Wesleyan University Press.

Marx, Karl. 1978. "On the Jewish Question." In *The Marx-Engels Reader*, ed. Robert C. Tucker, 26–52. New York: W. W. Norton.

Maus, Fred E. 2001. Narratology, Narrativity. In *The New Grove Dictionary of Music and Musicians*, ed. S. Sadie and J. Tyrrell, 641–43. London: Macmillan.

Mena, Jennifer. 2002. Iranian Anger on AM Dial. *Los Angeles Times*, December 29:B1, B14.

Merleau-Ponty, Maurice. 2002. *Phenomenology of Perception*. New York: Routledge.

Michaels, Walter Benn. 2006. *The Trouble with Diversity: How We Learned to Love Identity and Ignore Inequality*. New York: Metropolitan Books.

Mitchell, J. Clyde. 1956. *The Kalela Dance: Aspects of Social Relationships Among Urban Africans in Northern Rhodesia*. Manchester: Institute by the Manchester University Press.

Mitchell, Tony, ed. 2002. *Global Noise: Rap and Hip-Hop Outside the USA*. Middletown, Conn.: Wesleyan University Press.

Monson, Ingrid. 1996. *Saying Something: Jazz Improvisation and Interaction*. Chicago: University of Chicago Press.

Naficy, Hamid. 1993. *The Making of Exile Cultures: Iranian Television in Los Angeles*. Minneapolis: University of Minnesota Press.

Nancy, Jean-Luc. 2007. *Listening*. New York: Fordham University Press.

Neal, Mark Anthony. 2004. Postindustrial Soul: Black Popular Music at the Crossroads. In *That's the Joint! The Hip Hop Studies Reader*, ed. Murray Forman and Mark Anthony Neal, 363–87. New York: Routledge.

Negt, Oskar, and Alexander Kluge. 1993. *Public Sphere and Experience: Toward an Analysis of the Bourgeois and Proletarian Public Sphere*. Trans. Peter Labanyi, Jamie Daniel, and Assenka Oksiloff. Minneapolis: University of Minnesota Press.

Negus, Keith. 1999. *Music Genres and Corporate Cultures*. London: Routledge.

Nettl, Bruno, ed. 1978. *Eight Urban Musical Cultures: Tradition and Change*. Urbana: University of Illinois Press.

Nichols, Natalie. 2004. DaKAH's Fusion Fills Disney Hall. *Los Angeles Times*, March 16:E2.

Notley, Margaret. 1997. *Volksconcerte* in Vienna and Late Nineteenth-Century Ideology of the Symphony. *Journal of the American Musicological Society* 50(2–3):421–53.

Oboler, Suzanne, ed. 2006. *Latinos and Citizenship: The Dilemma of Belonging*. New York: Palgrave Macmillan.

Ong, Aihwa. 1996. Cultural Citizenship as Subject-Making: Immigrants Negotiate Racial and Cultural Boundaries in the United States. *Current Anthropology* 37(5):737–62.

———. 1999. *Flexible Citizenship: The Cultural Logics of Transnationality*. Durham, N.C.: Duke University Press.

———. 2003. Cyberpublics and Diaspora Politics Among Transnational Chinese. *Interventions* 5(1):82–100.

Oshun, Ifé. N.d. Los Angeles Hip-Hop Scene. Independent Artists Spotlight: Dakah. http://rap.about.com/library/bldakahinterview.htm (accessed April 14, 2005).

Paley, Aaron, and Aaron Slavin. 1989. *California Plaza Angel's Flight Productions, 1992–1996.* Los Angeles: Community Redevelopment Agency.

Park, Robert E., and Ernest W. Burgess, eds. 1925. *The City.* Chicago: University of Chicago Press.

Parson, Don. 1993. The Search for a Centre: The Recomposition of Race, Class, and Space in Los Angeles. *International Journal of Urban and Regional Research* 17(2):232–40.

Peterson, Marina. 2002. Performing the "People's Palace": Musical Performance and the Production of Space at the Chicago Cultural Center. *Space and Culture* 5(3):253–64.

———. 2005. Sounding the City: Public Concerts and Civic Belonging in Los Angeles. Ph.D. thesis, University of Chicago.

———. 2007. Translocal Civilities: Chinese Modern Dance at Downtown Los Angeles Public Concerts. In *Deciphering the Global: Its Scales, Spaces and Subjects*, ed. Saskia Sassen, 41–58. New York: Routledge.

Peyser, Joan. 1999. *To Boulez and Beyond: Music in Europe Since the* Rite of Spring. New York: Billboard Books.

PFUNK. 2003. Subjective Experience. http://www.PFUNK1.com/ (accessed November 4, 2003).

Polanyi, Karl. 2001 [1944]. *The Great Transformation: The Political and Economic Origins of Our Time.* Boston: Beacon Press.

Povinelli, Elizabeth A. 2003. *The Cunning of Recognition: Indigenous Alterities and the Making of Australian Multiculturalism.* Durham, N.C.: Duke University Press.

Prendergast, Christopher. 2000. *The Triangle of Representation.* New York: Columbia University Press.

Pugsley, William. 1977. *Bunker Hill: Last of the Lofty Mansions.* Pasadena, Calif.: Pentrex Media Group.

Qureshi, Regula. 2000. How Does Music Mean? Embodied Memories and the Politics of Affect in the Indian *Sarangi. American Ethnologist* 27(4):805–38.

Rabinow, Paul. 1995. *French Modern: Norms and Forms of the Social Environment.* Chicago: University of Chicago Press.

Rabinow, Paul, and George E. Marcus, with James Faubion and Tobias Rees. 2008. *Designs for an Anthropology of the Contemporary.* Durham, N.C.: Duke University Press.

Radano, Ronald M., and Philip V. Bohlman. 2000. *Music and the Racial Imagination.* Chicago: University of Chicago Press.

Rieff, David. 1992. *Los Angeles: Capital of the Third World.* New York: Touchstone.

Roach, Joseph. 1997. *Cities of the Dead: Circum-Atlantic Performance.* New York: Columbia University Press.

Robbins, Bruce, ed. 1993. *The Phantom Public Sphere.* Minneapolis: University of Minnesota Press.

Rogers, Alisdair. 2000. Citizenship, Multiculturalism, and the European City. In *A Companion to the City,* ed. Gary Bridge and Sophie Watson, 282–91. Oxford: Blackwell.

Rose, Tricia. 1994. *Black Noise: Rap Music and Black Culture in Contemporary America.* Hanover, N.H.: Wesleyan University Press.

Rousseau, Jean-Jacques. 1960. *Politics and the Arts: Letter to M. D'Alembert on the Theatre.* Trans. Allan Bloom. Ithaca, N.Y.: Cornell University Press.

Rubin, Rachel. 2004. Interview with El Vez. *Journal of Popular Music Studies* 16(2):213–20.

Russell, Paul. N.d. Hip-Hop's First Orchestra. http://www.rapnewsdirect.com/Artists/Black_Eyed_Peas/News/0-202-924-00.html (accessed November 4, 2003).

Ryan, Mary P. 1997. *Civic Wars: Democracy and Public Life in the American City During the Nineteenth Century.* Berkeley: University of California Press.

———. 2006. A Durable Centre of Urban Space: The Los Angeles Plaza. *Urban History* 33(3):457–83.

Samuels, David. 2004. The Rap on Rap: The "Black Music" That Isn't Either. In *That's the Joint! The Hip Hop Studies Reader,* ed. Murray Forman and Mark Anthony Neal, 147–53. New York: Routledge.

Sassen, Saskia. 1998. *Globalization and Its Discontents: Essays on the New Mobility of People and Money.* New York: Free Press.

———. 2000. *Cities in a World Economy.* 2nd ed. Thousand Oaks, Calif.: Pine Forge Press.

———. 2001. *The Global City: New York, London, Tokyo.* 2nd ed. Princeton, N.J.: Princeton University Press.

———. 2003. The Repositioning of Citizenship: Emergent Subjects and Spaces for Politics. *New Centennial Review* 3(2):41–66.

Saul, Scott. 2003. *Freedom Is, Freedom Ain't: Jazz and the Making of the Sixties.* Cambridge, Mass.: Harvard University Press.

Schein, Louisa. 2002. Mapping Hmong Media in Diasporic Space. In *Media Worlds: Anthropology on New Terrain,* ed. Faye D. Ginsburg, Lila Abu-Lughod, and Brian Larkin, 229–44. Berkeley: University of California Press.

Schloss, Joseph G. 2004. *Making Beats: The Art of Sample-Based Hip-Hop.* Middletown, Conn.: Wesleyan University Press.

Schmidt, Ronald J., Jr. 2005. *This Is the City: Making Model Citizens in Los Angeles.* Minneapolis: University of Minnesota Press.

Scott, Allen J. 2000. *The Cultural Economy of Cities: Essays on the Geography of Image-Producing Industries.* London: Sage.

Scott, Allen J., and Edward W. Soja, eds. 1996. *The City: Los Angeles and Urban Theory at the End of the Twentieth Century.* Berkeley: University of California Press.

Scott, Mel. 1949. *Metropolitan Los Angeles: One Community.* Los Angeles: Haynes Foundation.

Seeger, Anthony. 2004. *Why Suya Sing: A Musical Anthropology of an Amazonian People.* Urbana: University of Illinois Press.

Segal, Lewis. 2003. China Identified in Part and Whole: The Accomplished Beijing Modern Dance Company Is Finally Introduced to U.S. Audiences. *Los Angeles Times,* August 30. http://www.calendarlive.com/stage/segal/cl-et- (accessed January 13, 2004).

Self-Help Graphics: Tomás Benitez Talks to Harry Gamboa Jr. 2003. In *The Sons and Daughters of Los: Culture and Community in L.A.,* ed. David E. James, 195–96. Philadelphia: Temple University Press.

Shank, Barry. 1994. *Dissonant Identities: The Rock'n'Roll Scene in Austin, Texas.* Hanover, N.H.: Wesleyan University Press.

Shaw, William. 2002. *Westside: The Coast-to-Coast Explosion of Hip Hop.* New York: Cooper Square Press.

Shipley, Jesse. 2007. *Living the Hiplife.* New York: Third World Newsreel.

Small, Christopher. 1998. *Musicking: The Meanings of Performing and Listening.* Middletown, Conn.: Wesleyan University Press.

Smith, Catherine Parsons. 2007. *Making Music in Los Angeles: Transforming the Popular.* Berkeley: University of California Press.

Smith, Michael Peter. 1988. *City, State, and Market: The Political Economy of Urban Society.* New York: Blackwell.

———. 2001. *Transnational Urbanism: Locating Globalization.* Oxford: Blackwell.

Smith, Neil. 1996. *The New Urban Frontier: Gentrification and the Revanchist City.* London: Routledge.

Soja, Edward W. 1989. *Postmodern Geographies: The Reassertion of Space in Critical Social Theory.* London: Verso.

———. 1996. *Thirdspace: Journeys to Los Angeles and Other Real-and-Imagined Places.* Oxford: Blackwell.

Sonenshein, Raphael J. 1993. *Politics in Black and White.* Princeton, N.J.: Princeton University Press.

Sreberny, Annabelle. 2000. Media and Diasporic Consciousness: An Exploration Among Iranians in London. In *Ethnic Minorities and the Media,* ed. Simon Cottle, 179–96. Buckingham: Open University Press.

Starr, S. Frederick, ed. 1987. *The Oberlin Book of Bandstands.* Washington, D.C.: Preservation Press.

Stokes, Martin. 1992. *The Arabesk Debate: Music and Musicians in Modern Turkey.* Oxford: Oxford University Press.

———, ed. 1994. *Ethnicity, Identity and Music: The Musical Construction of Place.* Oxford: Berg.

———. 2000. East, West, and Arabesk. In *Western Music and Its Others: Difference, Representation, and Appropriation in Music,* ed. Georgina Born and David Hesmondhalgh, 213–33. Berkeley: University of California Press.

Strauss, Gloria B. 1977. Dance and Ideology in China, Past and Present: A Study of Ballet in the People's Republic. In *Asian and Pacific Dance: Selected Papers from the 1974 CORD-SEM Conference,* ed. Adrienne L. Kaeppler, Judy VanZile, and Carl Wolz, 19–54. Dance Research Annual 8. New York: Committee on Research in Dance.

Strohm, Reinhard. 1985. *Music in Late Medieval Bruges.* Oxford: Clarendon Press.

Tajbakhsh, Kian. 2001. *The Promise of the City: Space, Identity, and Politics in Contemporary Social Thought.* Berkeley: University of California Press.

Taylor, Charles. 1994. *Multiculturalism: Examining the Politics of Recognition.* Ed. Amy Gutman. Princeton, N.J.: Princeton University Press.

Thompson, Emily Ann. 2002. *Soundscape of Modernity: Architectural Acoustics and the Culture of Listening in America, 1900–1933.* Cambridge, Mass.: MIT Press.

Trouillot, Michel-Rolph. 1991. Anthropology and the Savage Slot: The Poetics and Politics of Otherness. In *Recapturing Anthropology: Working in the Present,* ed. Richard G. Fox, 17–44. Santa Fe, N.M.: School of American Research Press.

Turino, Thomas. 1999. Signs of Imagination, Identity, and Experience: A Peircian Semiotic Theory for Music. *Ethnomusicology* 43(2):221–55.

Vaillant, Derek. 2003. *Sounds of Reform: Progressivism & Music in Chicago, 1873–1935.* Chicago: University of Chicago Press.

Valle, Victor M., and Rodolfo D. Torres. 2000. *Latino Metropolis.* Minneapolis: University of Minnesota Press.

Walker, Victor Leo, II. 1989. The Politics of Art: A History of the Inner City Cultural Center, 1965–1986. Ph.D. thesis, University of California–Santa Barbara.

Wallenstein, Andrew. 2002. Many Tongues, One Goal: Asian TV Power. *Hollywood Reporter,* June 12:1–2.

Ward, Stephen V. 1998. *Selling Places: The Marketing and Promotion of Towns and Cities, 1850–2000.* London: E & FN Spon.

Warner, Michael. 2002. *Publics and Counterpublics.* New York: Zone Books.

Washburn, Jim. 1994. El Vez the Mexican Elvis: El Vez in 1994. *A Rockabilly Hall of Fame Presentation.* http://www.rockabillyhall.com/ElVez.html (accessed February 8, 2009).

Weber, Rachel. 2002. Extracting Value from the City: Neoliberalism and Urban Redevelopment. In *Spaces of Neoliberalism: Urban Restructuring in North America and Western Europe,* ed. Neil Brenner and Nik Theodore, 172–93. Oxford: Blackwell.

Weber, William. 2001. "Concert ii." In *The New Grove Dictionary of Music and Musicians,* ed. S. Sadie and J. Tyrrell, 221–35. London: Macmillan.

Whiteley, Sheila, Andy Bennett, and Stan Hawkins, eds. 2004. *Music, Space and Place: Popular Music and Cultural Identity.* Aldershot: Aldgate.

Whitt, J. Allen. 1987. Mozart in the Metropolis: The Arts Coalition and the Urban Growth Machine. *Urban Affairs Quarterly* 23(1):15–36.

Whyte, William H. 2001. *The Social Life of Small Urban Spaces.* New York: Project for Public Spaces, Inc.

Wild, Mark. 2005. *Street Meeting: Multiethnic Neighborhoods in Early Twentieth-Century Los Angeles.* Berkeley: University of California Press.

Wood, Denis. 1992. *The Power of Maps.* New York: Guilford Press.

Yang, Mina. 2008. *California Polyphony: Ethnic Voices, Musical Crossroads.* Urbana: University of Illinois Press.

Zukin, Sharon. 1982. Art in the Arms of Power: Market Relations and Collective Patronage in the Capitalist State. *Theory and Society* 11(4):423–51.

Index

www.ingramcontent.com/pod-product-compliance
Lightning Source LLC
Chambersburg PA
CBHW031135270326
41929CB00011B/1628